Ecology, Economy, Growth

Ecology
Economy
Growth

Chronicle of an unplanned journey

Melvin Bolton

Ribbonwood Publications

First published in 2019

Ribbonwood Publications
PO Box 879
Yeppoon
Queensland 4703
Australia

NATIONAL LIBRARY OF AUSTRALIA
A catalogue record for this book is available from the National Library of Australia

ISBN 978-0-9946142-1-6

Contents

Part three: Into the growth trap

Introduction

In a remote part of Papua New Guinea I once watched a woman using an adze to hollow out a log. The adze was a blade of stone attached in a traditional way to a wooden haft. The year was 1979, a decade after the first moon landing. We now live in a world of extreme contrasts but throughout history and prehistory there has never been a time of steady progression on a global scale. Change has always been erratic across time and place; the most important issues for one group having been irrelevant or unheard of elsewhere. And yet the most pressing problems of today are truly global and that makes them exceedingly difficult to tackle in a world where nations and regions are divided in so many ways.

The scope of this book is not restricted by conventional academic boundaries. It seems odd that most of what we learn in school or college is taught to us in separate subjects as if that is the way life is structured; this is geography, that is economics and there is biology. Knowledge is boundless and nothing could be more integrated than life itself, yet we have been teaching and learning by subjects since the middle of the nineteenth century and modern educational systems are based on it. Teachers, curricula, entrance requirements, certificates, diplomas and textbooks must all conform to the accepted academic disciplines. Not all educators are happy about it but the traditions are very strong, and whether we like it or not, the explosion of knowledge must be channelled into a structure of some sort. It is also obvious that advanced students must specialize in the areas of their chosen professions, and in one sense there is some truth in the joke about specialists having to know more and more about less and less. But there really is a downside to all this for society in general because the bigger picture is lost from view. And subject learning, together with the knowledge explosion and information overload, all combine to make bigger pictures even harder to see. Opinions that are formed on the strength of reports in the mass media are likely to be based on little snapshots or vignettes and that gives rise to arguments within a narrow context and without reference to the proper scale of the issues in question.

My aim is to provide a broad perspective from which to interpret the clamour about economic growth, environmental deterioration, the changing climate and the question of sustainability. I believe the big picture emerges most reliably from the narrative of what we are, where we have come from and how we arrived in our present circumstances. The story, based on available evidence, includes a number of critical turning-points that have occurred at very different rates in different parts of the world and are seldom considered within the covers of a single book. It is from this broader view that we gain new insights into how we came to our present predicament with regard to the natural environment, and why escaping from it is proving to be so difficult.

The internet, wonderful resource though it is, does not necessarily address the problem. We have become accustomed to finding information quickly and with no more inconvenience than thinking up a few good keywords, but in life the details that appear on the screen are all interconnected, and the connections are still easy to miss. In the case of ecology and economy the links are particularly strong because at the most basic level these two subjects are not separate at all.

Essentially, economics is the study of relationships between people and resources, whereas ecology is the study of relationships between all living things and resources. Modern economics has emerged from the business of managing the home estate whereas ecology extends the home estate to take in the whole environment, not just for humans but for all forms of life. The two words have the same Greek root: *oikos*, (pronounced 'eekos'), meaning 'house' or 'household'. But in the mass media the common roots of economy and ecology appear to be scarcely acknowledged and the two disciplines are associated with opposite ends of the political spectrum.

The earliest published works on economics date back to ancient Greece and were concerned with management at all levels, but the word 'ecology' was not used until 1866 and it has always meant more or less what it means today—the scientific study of interactions that determine the distribution and abundance of living things.

The three words 'interactions', 'distribution' and 'abundance' neatly distill the scope of the subject because what ecologists try to discover is why living things are in the places they are, and at the densities they are. Areas and numbers are at the core of ecology. The interactions of a species can be with other living things as well as with physical or chemical influences; all these combine to make up its environment.

Ecology, as a branch of biology, still has a long way to go; we don't even know approximately how many species there are, let alone how they interact. The subject has something in common with medicine and the social sciences insofar as it is unlikely to yield many principles that can be used to predict outcomes with great certainty. Biology and medicine can never be reduced to laws or formulae to the extent that we see in physics and chemistry because there are too many variables at work. Doctors can be sure what will kill a patient but there can be all sorts of reasons why two people will respond differently to possible cures.

Ecological studies can be focused on particular individuals, or populations of a species, or whole communities of many species, in each case including both living and non-living aspects of the environment. The word 'ecosystem' refers to all this at the community level. Ecology, then, as a branch of biological science, is not to be confused with environmentalism or green politics, although it may serve those interests just as scientific research can have a useful role in any social issue.

As for economics today, is it a science or an art? There is no simple answer. Definitions vary, some emphasizing different aspects of the subject, others reflecting the author's priorities. Regardless of the emphasis, economics is concerned with human beings and resources, and it has a lot to do with the way people make decisions about what they want. For that reason, if no other, economists also find it hard to make predictions with certainty. In general, as one Nobel laureate in economics pointed out, people want economic data as a guide to policy rather than something of interest for its own sake.[1] And when policy is the main interest we are soon involved in politics and ideology.

The Oxford dictionary defines economics as 'the branch of knowledge that deals with the production and distribution of wealth in theory and practice' and that leaves us needing some definition of wealth. A wealthy family in one society may have a few more cows or goats than the rest. Prior to farming, if we go back far enough, the wealthier group of humans might have had superior stone tools or were finding a better living in more supportive territory. Much the same can be said of the economies of chimpanzees or more lowly creatures.

It becomes evident that, at the simplest level, ecology and economy really are much the same thing; energy is the common currency of life, and life consists in capturing it and converting it to usable forms. But the concepts, even at that level, can be expressed in vocabularies that reflect some differences in the thinking of ecologists and economists, especially in regard to growth. To an ecologist, populations are made up of individuals, and a population that is neither increasing nor decreasing will be stable. To economists, concerned only with human beings, populations are markets made up of consumers, and a population that is neither increasing nor decreasing is likely to be described as stagnant, suggesting something that is not entirely healthy. I propose the word 'stagnant' as a shibboleth to distinguish economists from ecologists, though, as we shall see, there are other ways to tell them apart!

Part one: Forces of nature and powers of the gods

1 Primal growth and the single currency

Nowhere on the surface of the earth is known to be older than the Jack Hills of Western Australia. By mapping individual atoms of radioactive decay-products inside mineral grains, scientists have dated the rocks at 4.4 billion years.[1] Nobody can be sure exactly how long it was after that before life started ticking, but for billions of years to come it would exist only in the form of microbes—tiny single cells that could survive on a young planet that was still churning with chemicals and had no free oxygen.

Microbes in great abundance lived as packets of electrochemistry with nothing to be seen that we generally think of as activity, though that is not to say that early life was ever simple. Life's biochemistry worked then, as it does now, on the inside of a double layer of molecules called a plasma membrane. And existing cells, both free living and massed into tissues and bodily organs, still reproduce themselves by splitting in two.

To begin with, the microbes (bacteria and archaea[2]) became plentiful by extracting their energy from unlikely sources such as compounds of sulphur or iron in the sea. Their descendants still thrive today in places with little or no oxygen, although there are now other bacteria that do need oxygen and some can manage with or without it. Microbes today are astonishing in their different types of metabolism; they have colonized the human body to the extent that they greatly outnumber human cells but because of their small size they amount to no more than three per cent of our body weight.[3]

Life on earth is based on carbon in combination with five other main elements—hydrogen, nitrogen, oxygen, phosphorus and sulfur. Details of metabolism fill textbooks on biochemistry but, basically, none of it would be possible were it not for the fact that atoms of these elements can form chemical bonds by sharing electrons. Electrons are sub-atomic particles with a negative electrical charge and they occupy different energy levels depending on their position relative to the nucleus of the atom of which they are part. When a bond is broken, electrons are set free to move to a different electron acceptor, and they carry their potential energy with them as they move to form new bonds. In living cells, regardless of whether the electrons came from a sulphur compound or a ham sandwich, their travel ends with the manufacture of a substance known from its initials as ATP (adenosine triphosphate). Molecules of ATP serve as a universal coinage inside living things because some of its bonds are easily broken by cellular chemistry and that releases energy on demand.

Biochemists are keen to know how ATP could have been put together under the conditions that prevailed when life began. Deep sea hydrothermal vents known as 'black smokers' might well have acted as incubators and research is ongoing. Kick-starting life in a laboratory will be a wonderful achievement but of course it will demonstrate only how it could have happened; it cannot confirm how it did happen.

What can be said with certainty is that almost all existing life builds ATP in much the same way. To put it very simply, electrons are shuttled through a sequence of electron carriers on the inner surface of a plasma membrane (or a later version of it as I will soon explain). The electrons release energy at key points in the sequence and the energy is used to pump positively charged particles (protons) through the membrane to the outside. The build-up of protons on the outside of the membrane increases their concentration relative to the inside. It also creates an electrical potential across the membrane—positive outside, negative inside. Because of this electrochemical gradient, protons then flow back inwards through special chemical channels that build ATP. The procedure is known as chemiosmosis, a word coined by Peter Mitchell who, in 1978, won the Nobel Prize in Chemistry for his explanation of it. Professor Brian Cox, in his popular BBC television series *Wonders of Life*, made much of the importance of proton gradients not only for sustaining life but for getting it started in the first place.

The coming of oxygen

The act of breathing, getting air into the lungs, is commonly called 'respiration', but respiration at the cellular level is the work of electron transport chains and the making of ATP. It had to be done without oxygen for hundreds of millions of years but that began to change after some bacteria improved their metabolism by harnessing the properties of pigments. There are lots of pigments in nature and it is because their molecules absorb light very well that they look coloured and we can use them in dyes and paints. More importantly for life, light is a form of energy and photons of light that are absorbed can excite electrons in pigments to a higher energy level.

Bacteria that evolved organic pigments in their membranes had different ways of using the energy of light but it was a group of water-splitters that changed the world. A molecule of water (H_2O) has two hydrogen atoms bonded with one oxygen atom. By splitting off the hydrogen, the bacteria could combine it with carbon dioxide to build a carbohydrate. The freed oxygen atoms joined themselves in pairs to form oxygen gas (O_2) as a waste product. Few forms of life can now survive without it. This building of carbohydrates by using the energy of sunlight is now familiar to everybody as photosynthesis in plants and algae but it is still widespread among bacteria, and some types can manage to do it with very little light. A type of green sulphur bacterium was found to be using the dim light given off by a deep sea black smoker at a depth of over two kilometres.[4]

Water-splitting bacteria may have been releasing oxygen on earth for a very long time before it had much effect on life in general. Oxygen is chemically very active and quickly combines with other elements, so the oxygen set free by photosynthesis in the oceans was soaked up again, especially by dissolved iron to form iron oxides. But, eventually, the oxygen sinks became saturated and free oxygen began to accumulate both in air and water.

During the period between about 2.4 to 2.0 billion years ago the oxygen level in the air and oceans probably reached more than ten percent of its present level.[5] It was a toxic gas to those microbes that had never encountered oxygen before so many types may have died out. But bacteria have a high mutation rate and rapid reproduction so are highly adaptable organisms—as they continue to demonstrate by evolving resistance to the best antibiotics.

Oxygen was well worth getting accustomed to by every form of life because it is a superior electron acceptor in electron transport chains. Cells that can use oxygen for their respiration have a huge advantage over those that survive without it. The problem for microbes, though, is that even with oxygen, they cannot be anything other than small, single cells. There had to be other changes before life could really take off.

Life gets more complicated

The early water-splitting bacteria appear to have been ancestral to types that now make up a major division of bacteria called the cyanobacteria (Gr. *kyanos*, blue). They are often called 'blue-green algae' but that is biologically misleading. It would make more sense to say that algae and plants have cyanobacteria living inside them but that needs explaining and we haven't quite reached that point; I haven't entirely finished with microbes.

The surface area of microbes gets relatively smaller the bigger they grow. It is easy to calculate that if you double the length, breadth and height of a matchbox you will have four times the surface but eight times the volume. The rule applies to all compact shapes and the same principle applies to irregular shapes. This rule alone is enough to keep microbes small because all the needs of the cell have to be served through the area of the surface, and the surface can't possibly keep up with the work if demands for energy from the interior are increasing disproportionately. For organisms to grow bigger there had to be another way.

Bacteria and archaea, are not the only forms of life that exist as free-living single cells. Others include such organisms as amoebas, diatoms, and many types of algae. With an ordinary microscope we can see that these cells have something in common with cells from plant and animal tissues. They have a cell nucleus that appears as a darker blob in cells seen through a school microscope. It is actually a dense packet of coiled chromosomes—long threads of DNA that is the stuff of genes and spells out an organism's genetic code. Cells of bacteria and archaea have their chromosome as a single loop that is not contained in a nucleus.

When cell nuclei appear as nothing more than dense blobs in cells then something like cherry stones might come to mind, which explains why all forms of life except bacteria and archaea are collectively called eukaryotes (Gr. *eu*, good; *karuon*, kernel). The microbes, having no nuclei, are prokaryotes. Regrettably, they have no names that are more familiar.

So, humans are eukaryotes along with every other form of life that we can see around us, and a nucleus is not the only thing that our cells have in common. All eukaryote cells are packed with structures so small that details can be seen only with electron microscopes. The structures are collectively known as organelles.

In particular, organelles called mitochondria (sing. mitochondrion) are key to understanding how life escaped the energy limitations that prevailed for more than two billion years. Mitochondria are variously spherical or sausage-shaped and are wrapped in a double membrane. The inner membrane has numerous folds that project inwards, multiplying its surface area enormously. This specialized inner membrane works like the plasma membrane of a bacterium, carrying out the cell's respiration and manufacturing ATP. In fact, biologists are now satisfied that mitochondria were once free-living bacteria.

Mitochondria have kept enough of their own genes to be able to divide, grow or die independently of the cells in which they live, but the cell nucleus acts as a command centre and mitochondrial genes have gradually been taken over by chromosomes in the cell nucleus. These gene transfers are still happening. Bits of DNA can be set free from mitochondria when they die. The human genome project has revealed that there have been more than 350 transfers of mitochondrial DNA to the nucleus in humans.[6]

How many mitochondria are we talking about? It depends on the energy needs of the cell. Cells that are metabolically very active, such as liver cells or those in a germinating seed, have a much higher density of mitochondria than cells that are relatively dormant, such as skin cells or those in plant tissue that is not growing. There can be thousands of mitochondria in one liver or muscle cell. Above all, mitochondria are adaptable. They can change shape, adjust their numbers and membranes, and even move about the cell to deliver energy on demand.

For decades it was thought that a bacterial cell must have been swallowed up in some way by an early eukaryote but more recently an alternative view has been gaining ground. In this newer hypothesis, both organisms were prokaryotes when one first took up residence inside the other.[7,8] It was this merger that created eukaryotes and opened up all future possibilities in their evolution. Exactly how it happened is another question, but among existing bacteria one species does sometimes have cells of a smaller species thriving inside it, so it can happen.[9] By whatever means, the event must have occurred very early in the evolution of eukaryotes because every known type has, or had, mitochondria. A few have subsequently lost them except for some of the mitochondrial genes. So if ever there was such a thing as a eukaryote that never had mitochondria, it has vanished without trace.

A few organisms have also lost the ability to reproduce sexually, but they are very few. Sex is another property of eukaryotes that evolved in single cells. Prokaryotes can join up and pass genes from one to another, but only eukaryotes have the molecular machinery to reshuffle and combine genes from two individuals so that all offspring have a full set of genes derived from two parents.

Bigger and better

With energy from mitochondria, cells could become bigger and, more importantly, could do a lot more work as instructed by their DNA. An average eukaryote cell might be about ten times as big as the average bacterium but has hundreds of times more DNA and a lot more energy per gene. It's not that genes need much energy for their existence but there is no point in having genes without the energy to carry out their instructions—i.e. expressing the genes. Genes are codes written in DNA and they spell out recipes for the vast number of proteins that act as the workforce of heredity—doing the job of building cells as prescribed. Gene expression takes up about three quarters of a cell's energy budget.[10,11]

By around a billion years ago, there were colonies of eukaryote cells that could function as one individual—like sponges. Sponges are animals at their simplest. They consist of a few different types of cell with a chemical recognition system; the cells can slowly recombine if they are sieved apart.

When life became multicellular, entirely new avenues for evolution were opened up. Cells could specialize in masses to become tissues. A band of tissue around a jellyfish works as a muscle that makes the bell contract so the animal pulsates its way through the water. Better still, lumps of food can be contained in the bell long enough for specialized cells to pour out digestive juices and turn big lumps into something that cells can absorb.

Specialization and division of labour among cells made all kinds of new efficiencies possible. The electrical potential across cell membranes is the basis of a nervous system. Jellyfish have only a loose network of nerve cells with nothing like a control centre. They don't even have a head-end because they are organized in a circle, that is to say they are radially symmetrical, as are sea-urchins and starfish. Bodies organized with left and right sides and a head-end (bilateral symmetry) appear in the fossil record well over 600 million years ago, but another life-changing event occurred before then and it led to the evolution of algae and plants. I can return now to those blue-green algae which are more accurately called cyanobacteria.

Nobody can be sure when cyanobacteria first began to survive as organelles inside bigger cells but it happened long after cells acquired mitochondria. The cyanobacteria became chloroplasts, the organelles that contain chlorophyll in plants and algae. Like mitochondria, chloroplasts are enclosed by a double membrane and have their own circle of DNA, but they have also become dependent on genes in the nucleus so neither chloroplasts nor mitochondria can survive outside the eukaryote cell.

In the geological time-scale the period between about 540 to 490 million years ago is the Cambrian, and rocks from that period contain a spectacular variety of fossils. It is a phenomenon known as the Cambrian explosion but to what extent it was a result of rapid evolution, or particularly good conditions for fossilization, is debatable. The balance of opinion seems to be that at least the main body plans all existed before the Cambrian.

Certainly, by the end of the Cambrian Period, there had been an astonishing diversification of animal types, including some that were venturing out of water. It would be tens of millions of years later, however, before plants had left their algal ancestors behind and were poking up as little green stems on dry land. Flowers, like the pollinating insects with which they evolved, were not leaving their mark until about 130 million years ago, and grasslands together with the grazing ancestors of our domestic stock, are a lot more recent than that.

People in the mix

With a focus on climate change, there is intense scientific interest in the last time the world suddenly warmed up. It was an event known as the Palaeocene-Eocene Thermal Maximum (PETM) and it occurred some 56 million years ago when the planet was already a lot warmer than it is now, with little or no ice at the poles. Over a period of a few thousand years the sea surface temperature at high latitudes rose by about another 6° C with a slightly smaller rise on land at middle latitudes. Temperatures dropped back again within the following 200,000 years.[12]

Mammals during that time were represented by many, though not all, of the major groups that we see today. The fossil record indicates that one group was distinct in having grasping hands and feet, fingernails and toenails (not claws) and forward-facing eyes. Like squirrels, these creatures lived in trees, but unlike squirrels they may have caught insects by hand and searched for them by sight. They were early primates, and genetic studies suggest their separation had begun millions of years earlier.

Many primates had been and gone by the time our own evolutionary line appeared. Apes became distinct only within the last 20 million years and most of those disappeared as the climate became cooler and drier and forest gave way to grassland. But evolution plays catch-up with environmental change and there are three types living today that diverged from a common ancestor surprisingly recently.

The chimpanzee and human lines have probably been biologically separate, with no interbreeding, for no more than about eight million years at most.[13] Gorillas were adapted for a separate niche three or four million years earlier, which is why chimps are genetically closer to humans than they are to gorillas: we share over 98 per cent of our genes with chimps. Biologically, we are all members of the family Hominidae, subfamily Homininae and there is no everyday conversational name for the group; not many folk think of themselves as apes. In 1758, the founder of the biological classification scheme, Carl Linnaeus, gave us the name *Homo sapiens* 'wise man'.

Clearly, a few million years of evolutionary change has moved us a very long way apart from other hominins but it would be wrong to think that the human lineage has been anything like a straight line to the present. From a collection of fossil types found in Africa, nothing looking much like a human emerged until about 2.5 million years ago, and there were still more separations to come. Even as recently as 50,000 years ago there were at least four genetically distinct types of human still living. Whether they should be considered as different species or subspecies is arguable, but there was certainly some interbreeding because populations of existing people have genes from two of the types that have become extinct. We'll return to this in chapter 4.

So, people emerged very recently in evolutionary time and we emerged into a world of incredible biological complexity. Nobody knows how many species are now living. Eleven million (excluding prokaryotes) is the estimate adopted by Australian specialists. Insects are the biggest group, with vertebrates making up less than one per cent and flowering plants only about three per cent of the total.[14]

All this life still lives and dies at the cellular level where the cells are fuelled by energy trapped in chemical bonds of ATP and released to meet cellular needs. In the living cells of every species the schedule of death and renewal is laid down in genetic code that runs its course for as long as the renewals endure; humans, in effect, build a new liver every year or so. The code itself is passed on in sperms and eggs, the single cells that are specialized for reproduction, and sex is forever increasing the variations by shuffling and combining their DNA. The complexity of it all is staggering but not really surprising. The copying of DNA is not always perfect and mutations have brought change, minute by minute, in countless individuals over unimaginable lengths of time. Change can be good, bad or indifferent in terms of its influence on individual survival and population growth, and in the natural way of things, verdicts are constantly being handed down in the inexorable numbers game of natural selection—which keeps the tally.

2 Thinking in terms of energy

How did we end up the way we are? This is a question to which I will return in chapter 10 but the point I want to make now is that natural selection is the main mechanism of evolution and it works by perpetuating or eliminating genetic changes in a gene pool according to their advantage or disadvantage for a particular way of life. There are always compromises, otherwise we could expect to have night vision like owls and would be able to sprint like cheetahs. Those qualities would surely have been beneficial for early humans had they been metabolically cheap and not in conflict with other priorities.

The costs and benefits that were being weighed in the balance as our earliest ancestors were evolving are not entirely obvious because the circumstances that prevailed during that time are long gone. But some assumptions can be tested. The selective pressures that gradually led to a completely upright posture, for example, are not at all hard to explain. In the currency of ATP it is a lot more cost-effective to walk upright on two legs than to walk on feet and knuckles like a chimpanzee so the gradual changes in skeleton and muscles could have been paying off in energy terms even with no other advantages. It has been shown, from the oxygen consumed on treadmills, that human walking is about 75 per cent less costly than is knuckle walking for chimps.[1]

Of course, being upright does also confer the advantage of having both hands free to carry things. That is an invaluable asset, though we have no direct evidence for what was carried to begin with because our most distant ancestors were up on two legs long before they were making stone tools.

As social primates we also have high priorities centred on mental abilities, such as speech and recognizing faces. Human brains are well adapted in this regard, just as the brains of other animals have specializations that we don't possess.[2,3,4]

A bird of the crow family, for instance, has an amazing memory for places. Clark's nutcracker (*Nucifraga columbiana*) survives the North American winters by remembering where it has buried many thousands of little caches of pine seeds. Some of us have trouble finding one car in an underground car park but this bird's spatial memory is phenomenal; it can even find its seeds after snow has fallen. The part of the brain called the hippocampus is proportionally bigger in this species and in others with similar abilities. In one fascinating piece of research it was found that when London taxi drivers had learned their way around London well enough to pass the necessary licensing test (a feat they call 'getting the knowledge') they were found to have a larger hippocampus.[5]

We can only make informed guesses about the advantages of big brains in early humans and we must be cautious with our assumptions. On the cost side, we do know that brains are metabolically expensive organs to maintain and they increase head size, which adds to the problems when mothers are giving birth. The first member of our genus, *Homo habilis*, walked the Great Rift Valley with a brain of a little more than 600 cubic centimetres compared with an average of 1,200–1,400 cc for the anatomically modern humans that emerged within the past 300,000 years. The human brain has not increased in size during that time; it amounts to about two per cent of body weight but uses twenty percent of the energy budget.[6]

While nobody can be sure why bigger brains evolved when they did, it is likely that there was a combination of reasons rather than a single cause. But language does seem to have been central to it. Human language is unique because it can put any thoughts into words—even original thoughts and nonsense rhymes. Other animals have some highly effective methods of communication but the type and quantity of information being expressed is always limited, not open-ended.

Obviously a brain must have produced the thoughts that are spoken, so the processing centres for speech and thinking must have expanded in tandem. But mutations would offer some advantage in this connection only if there were other individuals who were responsive to the changes, so the size of social groups may also have been important.

Not surprisingly, primates that live in bigger social groups have been found to have a bigger neocortex—the most recent part of the brain that overlies the cerebral hemispheres. In most mammals it is convoluted to increase its surface area, but in humans the process has gone furthest and the neocortex lies over most of the brain's upper surface so that, when seen from above, the whole brain looks like a soft grey walnut.

Primates show a very wide range of group sizes from the semi-solitary orang-utan to mandrills that assemble in hordes of hundreds, but group size within a species varies with changing ecological conditions. Sub-groups can form or rejoin depending on what works best in foraging for the same resources at the same time. All mammals know what they want and social harmony can soon break down—especially in times of scarcity and competition.[7]

In a research project by Robin Dunbar, a British primatologist, the correlations between neocortex size and average social group size were collected for 38 genera of primates. The findings suggested that in favourable ecological circumstances the maximum size of a cohesive, stable group will be determined by the 'social intellect' provided by neurons of the neocortex. It is not hard to see how language would prove highly beneficial in that regard.

The same findings, extrapolated to humans, suggest that we should be able to maintain close, stable relationships with somewhere between 100 and 230 individuals. Two decades ago the figure of 150 was adopted by the popular press and publicized as 'Dunbar's number'. It has more recently been applied to social networks online. Dunbar pointed out that 150 is about the splitting point of Hutterite colonies and the estimated size of a neolithic farming village.[8] But while there must be limits, there is evidently plenty of flexibility in the effective size of groups. The work of some anthropologists indicates that bands of hunter-gatherers probably averaged around 30 individuals and they ranged over large areas, probably covering 175 square kilometres even in warm, rich environments.[9] To get some idea of the scale of this, think of a circle with a settlement in the middle; the foraging distance around the settlement would be about 7.5 km. This is not to suggest that territories are ever likely to be circular except in very monotonous country; it is more realistic to think of territory in terms of 'sites and paths within a landscape' rather than compact areas.[10]

Families or bands in rich habitats could have exploited the surroundings within a day's walk of a settlement for as long as the rewards were worth the effort. In less productive regions they could have been more or less nomadic and camped in a different place every night to spread the foraging impact. They might have used different seasonal strategies. Studies of hunter-gatherer societies suggest that in general there was plenty of leisure time and foragers could perhaps collect as much as ten times the energy in food that they used up in collecting it. This concept of energy return on investment (EROI) has wide application in modern society and we shall return to it.

So far in this chapter I have discussed energy costs only in relation to individuals and organs but all levels of organization, from cells to ecosystems, are configured by the availability of energy. This is crucially important in the context of food and agriculture, but for the moment, think of a simple food chain: a leaf is eaten by a caterpillar, which is eaten by a bird, which is eaten by a cat. The leaf is the primary producer but only a proportion of the energy that went into it is obtained by the caterpillar. Similarly, only a proportion of the energy needed to grow the caterpillar is acquired by the bird, and the cat gets a small proportion of that.

After an animal has stopped growing, or getting fatter, its energy goes into maintenance so nothing further is passed on to a consumer higher up the food chain. And not all animals convert the same percentage of energy into tissue; some creatures need a higher proportion than others just to keep warm and stay alive. Actual figures vary widely but throughout an ecosystem it is realistic to think of a ten per cent transfer of energy at each link in a food chain. This fact alone could probably explain, theoretically, why food chains usually have no more than three or four links, and rarely more than six. The energy fixed by the primary producer will not stretch any further. In reality, the chains are always intersected by others to form webs; the caterpillar might have been eaten by a lizard, which could have been grabbed by a bird or a snake. Food webs in the sea can include hundreds of species with thousands of links.

Energy and human populations

Recently it has been suggested that the populations of some hunter-gatherers had begun to expand before agriculture, perhaps as a result of improved hunting technologies.[11] Quite possibly the inhabitants of some regions were more efficient exploiters than those in other places, and opportunities would not have been the same everywhere. For tens of thousands of years, for example, people in Africa, Australia and elsewhere are known to have practiced a sort of proto agriculture by burning vegetation and favouring species that were useful to them, but nonetheless they were being supported by natural ecosystems. To put it another way, they were living from day to day as part of a local food web so any population expansion would be limited by the natural carrying capacity of the habitat.

As a consequence, population densities were low. An analysis that combined archaeological data with climatic and ecological variables indicated that the population of Europe, before agriculture, would have averaged about 30,000.[12] That's roughly three times the current population of wolves in Europe.[13]

In the longer term, entire ecosystems come and go with the climate but within a human lifetime there are good times and lean times and this would have had consequences for population growth that are not always appreciated in the literature. So here we need to think a little about the way that populations are regulated.

All animal populations are ultimately dependent upon vegetation, which in turn depends on water, warmth and soil nutrients. As long as conditions are optimal, animal populations grow exponentially, which is to say that the offspring are added to the breeding stock in the way that interest is added to the principal by investors wanting compound interest. The 'rule of 70' applies in both cases: the doubling time can be found by dividing the annual rate of increase into 70, so at five per cent the population (or investment) will double in fourteen years. The rule doesn't work so well at very low or very high rates but is quite accurate up to ten per cent.

Under ideal conditions the rate that can realistically be achieved by animal populations depends on the type of life history. At one end of the spectrum there are some fish and insects that produce huge numbers of offspring and have nothing further to do with them, and at the other extreme there are animals such as elephants and humans that produce relatively few babies but spend years caring for them.

The maternal instinct can be very strong in some animals but elephants can have no more understanding than fish as to what it is that produces babies; only people know that. No other animal can have learned that copulation causes pregnancy and makes babies arrive after a delay of several weeks or months. We couldn't teach the cleverest dog a simple trick when cause and uncertain consequence are that far apart. So only humans can possibly understand what sex is all about, and until very recently people assumed that some mysterious power other than copulation must also be involved—hence all the fertility rituals.

Getting back to growth rates, we can get some insight into the maximum rate of increase for humans from studies of isolated populations such as the Hutterite communities in North America. The Hutterites did not marry until they were twenty, yet between 1880 and 1950 the population grew without immigration from 443 to 8,542 individuals—an annual rate of increase of 4.12 per cent. That's a doubling time of slightly under 17 years by the rule of 70. The authors of the study estimated the maximum fertility for humans to be twelve to fourteen children.[14] The word 'fertility' can lead to confusion. It is used here by demographers to mean what biologists would call 'fecundity'. When biologists talk of fertility they refer to the quality of being fertile, that is able to reproduce, regardless of whether reproduction takes place.

A population that is growing as fast as it can is said to be reaching its biotic potential, which depends on the survival of offspring as well as the fertility of parents. Life expectancy at birth is used by demographers as a measure of survival. It is determined both by the biology of the species (its possible life span) and environmental conditions that affect survival. During different prehistoric periods the life expectancy at birth could have ranged between 20 and 40 years but that is not to say there were no older people. Life expectancy at birth is based on averages and is the number of years that a newborn can be expected live, so for example, if infant mortality is very high then a two-year-old will have a higher life expectancy than a newborn. Up to a point, and with good health care, the longer you survive, the longer your life expectancy is likely to be.

In nature, most environments are not constant for very long periods and animal populations respond to the changes. There are many different scenarios but in general there is a dynamic relationship between food and feeders. When feeders begin to outgrow food supplies they suffer the consequences. Inadequate nutrition will not support biotic potential so failing food supply is reflected in slower population growth. As hunger becomes starvation there is a rise in mortality, especially among the very young and old. In response to the reduced population size, food supplies might recover if other conditions permit it. Populations of feeders can then begin to grow again. This is all very obvious to livestock managers who need to prevent the size of their herds from exceeding carrying capacity of the pasture. They hate to see their animals losing condition.

In population ecology, carrying capacity is denoted by k. It is a theoretical value that represents the population density that can just be maintained by available resources. With reference to k, populations show different patterns of growth—even within a single class such as mammals. The populations of some types with a high biotic potential, like rodents, can overshoot k for a short time, then crash below it to some extent only to recover and climb again to another overshoot. On a graph, their populations zigzag above and below k in a series of peaks and crashes. Populations of other types are more finely tuned to environmental conditions so the graph shows smaller fluctuations and is somewhat closer to a wavy line around k. Humans are in this category.

In the language of economics, k can be referred to as the subsistence income or subsistence wage, being the material income at which birth rates equal death rates in human populations.[15] As we shall see in later chapters, humans slowly increased k for their own populations by finding ways to extract more energy from their lands. Better tools, weapons and traps, cooking and control of fire presumably all helped in this regard but still populations grew very slowly until relatively recent times. Carrying capacity was increased in a spectacular way when lands were converted for agriculture, food was stored, and machines were driven by energy released from fossil fuels.

The mechanism for population control in response to carrying capacity (or subsistence income) is to be found in the combination of timing, nutrition and energy in females. Human gestation is more or less fixed at nine months but the timing of other important events, controlled by hormones, is far more variable because the endocrine system is highly responsive to nutrition and available energy. It is common knowledge that girls begin menstruation (menarche) earlier in wealthy societies than in poor ones—mainly varying from the thirteenth to the sixteenth year. But for population growth this is less significant than is the spacing of births during reproductive life.

When women begin breast-feeding they do not menstruate and are not fully fertile again for some time. The length of that time can be anywhere from a few weeks in affluent societies to two years or more among !Kung women of the Kalahari—who are breast-feeding for as long as there is any demand from the infant. For a woman, metabolizing for two is quite enough because milk production generally uses 15 to 25 per cent of her body's energy budget. For a woman in very poor circumstances it can be up to 50 per cent. Moreover, the pituitary hormone, prolactin, gives a high metabolic priority to milk production so demand will be met from a mother's fat reserves before milk supply begins to decline. When nutrition is poor and women have little or no reserves of body fat, then populations grow even more slowly, or not at all, as a direct result of infant mortality. Intervals of less than two years between births increase the risks significantly.

Early research appeared to indicate that more frequent breast-feeding extended the time of infertility following a birth and it did so independently of the total volume of milk expressed. A more recent interpretation is that available energy is what makes the difference, just as it affects the probability of conceiving. When a higher proportion of a mother's metabolic budget is devoted to milk production she will remain infertile for a longer period after she has given birth, regardless of how frequently she breast-feeds.[16]

Given the opportunity, we can easily control what we eat but we can't tell our molecular machinery exactly what to do with the energy; it can be hard enough for some of us just to lose or gain a little weight. During the 1960s in the Gambia, seasonal food shortage was leaving some villagers seriously undernourished. Nutritional supplements were given to pregnant and lactating mothers in order to improve infant health and survival prospects. The programme achieved a 45 per cent increase in mothers' energy intake but birth weights increased by only about 50 grams. Average milk output remained the same with only a small improvement in its quality. Milk, it seems, had already been a high priority in the mothers' metabolism. On the other hand, with better nutrition, prolactin levels declined much more quickly after birth, menstruation recommenced earlier and the interval between births shortened by nearly a third. The benefits went into population growth.[17] And that brings me back to populations.

Within a given home range the share of resources available for each individual depends to a large extent on the population density. Of course there would have been cultural responses among pre-agricultural people; cooperation, competition and territoriality all affect social spacing. Deliberate abortion and infanticide might possibly have been widespread in the distant past; both practices having been recorded among Yanomami people of Venezuela, who were still living largely as hunter-gatherers in the 1960s.[18] This would not alter the conclusion that prehistoric populations were characterized by high mortality sustained by natural fertility.[19,20]

In general, whatever the mix of mechanisms in a particular culture, as population density increases towards k there will be negative feedback in terms of available energy and this works to slow population growth. If the population falls below k, then the reduced pressure on resources may allow forage to recover so that negative feedback will cease and populations can begin to grow again. In other words, mortality and population growth are often density-dependent. Of course, if the decline has an independent cause, such as drought or cold, then food supplies may not recover as a result of population decline; the carrying capacity will remain lower until the drought breaks or the earth warms up. The resource base is always the bottom line and in the long term that largely depends on climate. We'll look at that in the next chapter.

3 Climate

It is often said that climate is what you expect, and weather is what you get; the difference between the two being only a matter of the recorded time. At any one place on earth the weather might change from day to day, or within the hour, but the climate for that place is an average of the weather elements (temperature, humidity etc.) over a long period as recorded in 30-year stretches known as 'climate normals'.

Climate basics

Most of what we experience as climate is a result of the warming of continents and oceans by the sun, and although the sun's radiation is not entirely constant, it varies only slightly on a cycle that averages about 11 years. Of course, neither the earth's surface nor the atmosphere could keep storing the sun's energy without getting hotter and hotter, so for the earth to stay more or less in equilibrium the incoming energy must eventually be re-radiated back into space. That would not be so complicated if the earth were just a black ball, but for a spinning planet with vast oceans around great land masses and high mountains, all surrounded by an atmosphere of different gases, it is all very complicated indeed.

Things that affect climate are known as forcing mechanisms and there are two categories of them. First, there are the external mechanisms of solar radiation and movements of the earth relative to the sun, and secondly there are all the internal mechanisms resulting from features of the earth itself and its atmosphere. Broadly, climates result from the ways in which sunshine is converted to different forms of energy and distributed around the earth in response to these forcing mechanisms.

External forcings

The earth makes one orbit round the sun each year, and because the earth is tilted, the northern hemisphere will be tilted towards the sun for half of the orbit (northern summer, southern winter) and tilted away from the sun on the other half (northern winter, southern summer). The orbit is not quite a circle, and its eccentricity varies from being at its most elliptical to its most circular on a cycle of about 100,000 years. What's more, the earth very slowly changes its angle of tilt. The tilt (obliquity) is currently 23.5 degrees but it varies by nearly two and a half degrees over a cycle of about 40,000 years. And finally, the earth is subject to 'wobble' as it spins, rather in the way that a spinning top can wobble as it slows. The technical word is 'precession' and the precessional cycle is roughly 22,000 years.

These three external forcing mechanisms, eccentricity, obliquity and precession, are known as the Milankovitch cycles after the Serbian mathematician Milutin Milankovitch (1879–1958) who did an impressive job of working them out without so much as an electronic calculator. They are now standard textbook fare.

Without getting into details, it's easy to see that a more eccentric orbit that brings the earth closer to the sun at the nearest point will also increase earth's solar radiation at that time. An increased angle of tilt causes more extreme seasons—hotter summers and colder winters—in both northern and southern hemispheres.

The effect of precession is different. Regardless of the angle of the axis as the earth spins, the wobble causes the north and south poles to move round in circles and that determines the angle at which a hemisphere is presented to the sun as the earth travels round its orbit. If a hemisphere is most directly facing the sun when the earth comes closest to it (perihelion), then that same hemisphere will be facing the opposite direction when the sun is furthest away (aphelion) and the seasons will be more extreme for that hemisphere. The seasonal difference will be less extreme for the other hemisphere.

The cycles of these external forcings are so long-term that for thousands of years at a time the radiation arriving at the earth's outer atmosphere hardly varies—which is why the measured amount is known as the solar constant. What happens to it during the period between entering the atmosphere and its being eventually re-radiated into space is anything but constant because that is where all the internal forcings and dynamics come into play.

Internal forcings

Not all the incoming radiation actually reaches the earth's surface; some of it is reflected away or selectively scattered as it passes through the atmosphere. A small proportion of that which does reach the surface is reflected back from ice and snow. Water, surprisingly, does not reflect very much at all.

In total, slightly more than half of the incoming radiation is absorbed at the earth's surface, and nearly half of that consists of visible light.[1] The shorter (higher energy) wavelengths from the ultraviolet end of the spectrum are mainly filtered out, although the small amount of UV light that does reach us is enough to cause skin cancer. Ozone molecules are particularly effective in absorbing UV rays. Ozone is a form of oxygen with molecules of three atoms (O_3) instead of two. It occurs naturally as a rare gas in a band of the stratosphere about 20 km thick.

Other internal forcings include plate tectonics (which move the continents and cause volcanic eruptions), currents of the oceans, and life itself—which is key to the carbon and nitrogen cycles as well as the level of oxygen in the air. There are multiple interactions and feedback responses between these forcings, some of which are fairly obvious and predictable; others much less so. Until very recently, for instance, it was thought that oxygen levels could not affect climate, but computer modelling suggests that over geological time it would have altered the amount of short wave scattering in the atmosphere and been an important influence on surface temperature.[2]

We talk and wonder about whether other planets might sustain life, but life is such an integral part of earth that it makes sense for us to ask to what extent life sustains the planet. Without life, earth would have a different atmosphere, a different climate, different surface minerals and no soil. The way our planet functions is the result of all the interactions between its living and non-living components. In that sense it works like a living body, like a gigantic, self-regulating superorganism. James Lovelock popularized the name 'Gaia', for the planet, and has been hugely influential in generating a more holistic view of the world, both among scientists and the general public.[3] Earth system scientists may stop well short of describing the earth as a living organism but the idea works extremely well as a metaphor, and Lovelock is too well-qualified for his ideas to be dismissed as uninformed opinion.[4] Gaia was the earth goddess of ancient Greece, and the name was actually suggested to Lovelock by his literary friend and neighbour, William Golding.

Upon reaching earth's surface, nearly all the radiant energy is converted to heat, which then warms up the continents and oceans. Only a tiny proportion of the energy is used in photosynthesis. A good crop of plants might build plant tissue from five per cent of the sunshine received, but the percentage averaged over the earth's surface works out at less than half of one per cent.[5]

Slightly more than 70 per cent of the earth is under water, and water can hold a lot of heat without much rise in temperature, which is to say that it has a high specific heat capacity. This is obvious when you put your hand on a rock, then in a puddle, on a hot day. To put it another way, for the same rise in temperature, water holds several times as much heat as does soil or rock so the oceans are the world's great reservoirs of heat. And when water evaporates from their surfaces, the vapour holds the heat of vaporization and does not release it until the vapour cools back to water. Water may fall as rain a thousand kilometres distant so water vapour is the atmosphere's main transporter of heat.

Water vapour is also by far the most important of the so-called greenhouse gases that delay the re-radiation of heat back into space. Without the greenhouse effect the planet would never get warm enough to have liquid water, which is essential for life. But the amount of water vapour held in the atmosphere depends on temperature, and the temperature is raised by quite small increases of other greenhouse gases, notably carbon dioxide, methane, nitrous oxide and ozone. An increased temperature allows the atmosphere to hold more water vapour, which further raises the temperature, and so on in a positive feedback effect that is creating a more effective heat trap than is good for us. It is a delicate balance.

So the warming of continents and oceans by the sun is the engine that powers the climate, and its most basic pattern around the earth can also be explained with familiar science. Around the equator the surface warms up most quickly as it is most nearly at right angles to the sun. ,The warm, humid air rising from the oceans creates the equatorial low pressure belt, which pulls in the trade winds, that are angled as they are because of the spin of the earth. The most basic facts really are quite easy to understand.

The basics are less obvious below water. Wind moves water in surface currents down to three or four hundred metres but deep ocean currents are generated by changes in density and salinity, which result from changes in temperature. It is a peculiarity of water that it becomes more dense as it cools to 3.98° C, but then it gets less dense—which is why ice floats.

When seawater freezes, the ice crystals leave out the salt, so compared to the ocean depths, the water immediately beneath polar ice is not only more dense because it is colder, it is also more dense because it is saltier; it holds the salt displaced from the ice above. This causes surface water under polar ice to sink slowly into the deep, displacing phenomenal volumes of less dense water as it does so, and maintaining a temperature-salt (thermohaline) current that travels round the world.

Think of this ocean conveyor belt as beginning off Norway, moving south through the western Atlantic as far as Antarctica, where it gets recharged beneath the ice, then moving north through the Indian and Pacific oceans until the currents begin to circulate as warmer water. The process is continuous but it would take centuries for a particular unit of water to complete the circuit.[6] On a day-to-day basis, as well as affecting climate directly, the conveyor belt raises nutrients to depleted surface waters and so supports the algae that make up the base of all marine food webs—which in turn are an integral part of the carbon cycle. The ocean conveyor could scarcely be more important for the earth's ecology but it is poorly understood. Recent scientific monitoring suggests that the Atlantic part of it is slowing and weakening.[7,8]

Finally, there is the ever-present chance of a major volcanic eruption. In geological time the shifting of the earth's tectonic plates builds mountains and moves entire continents about, but the climate can be affected for months or years by the activity of a volcano during a single day. Like earthquakes, most volcanoes are associated with the edges of tectonic plates, though some may melt through plates and erupt far from the margins as a result of what are thought to be hot spots in the magma beneath the crust. Volcanic eruptions are spectacular but the heat and clouds of ash have not been significant for the climate in recent years; ash has generally cleared within weeks. On the other hand, sulphur dioxide is emitted in megatonnes and it rises to the stratosphere where it is dispersed and converted to aerosols of sulphuric acid. Aerosols, tiny suspended droplets, reflect sunshine back from the atmosphere so have a cooling effect.[9]

In view of such massive controls on climate, one might wonder why it is so variable in the short term, but variability is not really surprising in view of all the interactions and feedback effects. Ice ages, for instance, are very long periods when ice sheets extend well beyond the polar regions. Within an ice age, on a scale of tens of thousands of years, there are glacial and interglacial periods when the ice expands and retreats, and within a glacial there are minor advances and retreats of the ice that are known as stadials (colder) and interstadials (warmer). The Milankovitch cycles have clearly been a major influence on the timing of ice ages but there are anomalies that have yet to be explained.[10]

By comparison, a year-to-year phenomenon known as El Niño is quite predictable. It occurs when surface water off South America becomes warmer than average and extends along the equator to join up with warm water off Australia. Gradients in sea surface temperature create pressure gradients that in turn direct winds. The end results of El Niño tend to be drought in Australia, heavy rain in parts of South America, and knock-on effects at higher latitudes. The opposite of this situation is La Niña, when cool water off South America is more extensive than usual.

But even if all the known complexities were to be fed into the most sophisticated computer models, there would still remain the unpredictable. The question whether a butterfly's wing-beat in Brazil could lead to a tornado in Texas was first asked in 1972 and people have been wondering about 'the butterfly effect' ever since. It was a question that made a point. The fluttering of butterflies might never be important but tiny events in large numbers, each individually insignificant and beneath the precision of our measurements, might not be entirely random and could conceivably work together to tip the balance of something somewhere. This is the realm of an arcane branch of mathematics called 'deterministic chaos' theory.[11] I shall leave it there!

4 People through the Pleistocene

Not everybody agrees on the best way to define a modern human. Until very recently, it was thought that the oldest fossils of modern humans were those recovered from the Omo River region of south-west Ethiopia and dated at 195,000 years old, give or take a few thousand years.[1] Fossils from a site in Morocco, however, have now been dated at 300,000 years, and their mix of anatomical features suggests that the evolution of our species may have taken place over a longer period, and over a wider expanse of Africa, than was thought.[2]

But some scientists attach more importance to intellect and behaviour than to anatomy. The problem is that we can only infer intellect and behaviour from the tools, artifacts and similar hard evidence that has come to light, and that increases the problem of deciding which human types to call 'modern'. Fortunately, the problem is not central to the purpose of this book.

Regardless of their brain power and behaviour, all humans (genus *Homo*) evolved entirely within the Pleistocene epoch—a time of dramatic changes in climate that affected the entire planet. The Pleistocene epoch has been variously defined but by international agreement it is the time between about 2.58 million years ago and the start of the Holocene, only 11,700 years before the present.[3]

The beginning of the Pleistocene marks the start of the most recent ice age, and during its glacials much of the northern hemisphere was under ice. Ice sheets extended across Europe from the British Isles to Mongolia and south into the highlands of central Europe. In Norway, parts of the Scandinavian ice sheet were possibly more than three kilometres thick though the average thickness was more likely to be a little less than two kilometres. Further south, glaciers formed in the mountains around the Mediterranean, in Turkey, Greece, the Italian Apennines, the Pyrenees and the High Atlas mountains of North Africa.

Perhaps more importantly for human existence, there was only tundra vegetation for a long way south of the great ice sheets, and this would have been underlain with frozen subsoil over very wide areas. The patterns left by this permafrost can be detected in aerial photographs and it is thought that the only part of mainland Britain not affected by it would have been in the area that is now Cornwall.[4] Similar conditions prevailed in North America, with most of Canada under ice that stretched as far south as Illinois in the east. At their thickest parts the ice sheets were over three kilometres thick—more than heavy enough to depress the Earth's crust. During glaciations water remained locked up in ice rather than flowing back to sea so global sea levels were tens of metres, and sometimes more than 100 metres, lower than they are today.[5]

In Africa the periods of glaciation were not only colder, they were also much dryer. In addition, the Great Rift Valley, with its chain of lakes, was affected by tectonics and rifting with relatively rapid changes between very wet and very dry periods during which the lake levels fluctuated enormously. It has been suggested that the effects of these climatic extremes would have extended in Africa far beyond the rift system.[6] We should bear in mind, though, that when geologists say 'rapid' they are usually talking about periods that span tens of human generations.

Becoming human

Among the earliest Pleistocene fossils the animal most resembling a human being stood about 1.3 metres tall and had a brain volume of around 600 cc, which is bigger than that of any chimpanzee. The first fossils were discovered in the 1960s and were eventually assigned to our genus as *Homo habilis* (handyman), which has become the widely accepted name, although not all specialists agree with the classification. Some see the *habilis* types as having more in common with *Australopithecus*, an earlier genus that became extinct during the Pleistocene. Others are not convinced that *habilis* belongs to either genus.[7]

Nor was the name 'handyman' conferred with total confidence but the litter of stone tools at the fossil site had certainly been handmade by something living around two million years ago, and although other *Homo* species may have existed at the time, no better candidate has been found. Simple tools of wood and bone were presumably among the first ever used but have left no evidence from these earliest times. The stone tools resemble cobbles with rounded channels and cavities where flakes have been chipped away. It is now believed that the sharp-edged flakes, a few centimetres long, were the actual cutting and scraping tools and they were banged off the cobble with a hammer stone.

Stone tools of this type are examples of the Olduwan industry, named after Olduvai Gorge in Tanzania where *H. habilis* was first found. In Africa, similar tools have since been discovered at a number of sites between South Africa and Ethiopia and, although they show some variation, they demonstrate that the early hominins, generation after generation for nearly a million years, were making their tools by knocking two stones together in much the same way. Simple as this may sound, it is not a knack that could be acquired by a captive chimpanzee (actually a bonobo, *Pan paniscus*) although to be fair on chimps, they do have less flexible wrists as well as smaller brains.[8,9]

There is no way of knowing whether *H. habilis* contributed to our own stream of DNA, or whether its type came and went like so many other species whose genetic tributary simply dried up. Fossil finds have been dated between about 2.4–1.6 million years old.

We are better informed about *Homo erectus*, a type that evidently coexisted for a time with *H. habilis* because the oldest fossils in Africa are dated at about 1.8 million years, and finds of both types have been made at the same site near Lake Turkana in Kenya. The most complete *erectus* skeleton was unearthed there in 1984. It was the fossilized remains of a boy who died some 1.5 million years ago.[10]

The Turkana boy had walked fully upright on long legs and was indisputably better adapted for life in open savannahs than among dense trees. His kind had a brain size of about 850–900 cc and the chewing teeth were smaller than those of earlier forms, indicating a softer diet, perhaps with more meat. But this is not to suggest that *erectus* was a biologically more specialized hunter, though early humans may have become more adept at both hunting and scavenging according to some who have studied the marks of carnivore teeth and stone tools on fossil bones. Different tactics were probably used in different seasons and habitats as is the case with many other animals. There seems to be no doubt that both hunting and scavenging took place.[11]

Anatomically, human beings are generalists, able to eat and digest just about any other form of life—from fungi and snails to other primates. In ecological terms we have always occupied a broad 'fundamental niche'. The extent to which that niche has been realized has changed along with culture and technology rather than biological evolution. Cooking, in particular, would have made a big difference. Hearths became common in the archaeological record from around 400,000 years ago but clear evidence of cooking has been found in a cave in South Africa and dated at about one million years ago.[12]

The early types of *erectus* showed no sudden advances in stone technology. Tools continued to be made more or less in the Olduwan tradition for thousands of generations after the time of Turkana boy. But another tradition, the Acheulean, was already developing in which stones were being chipped into a shape that was already in the mind's eye or was there to be copied.[13] For a million years or so the Acheulean tradition would produce tools that specialists have classified and graded according to design and standard of technology. The most commonly depicted type is a hand-sized lump of rock, roughly pear-shaped but flattish, with a cutting edge narrowed and sharpened by chipping from both sides. This is described as a hand axe and its particular form in any period varies according to locality, type of stone, usage and resharpening.[14] The name comes from St-Acheul, a suburb of Amiens in France, where tools of this sort were discovered in the nineteenth century.

It would be satisfying to be able to chart a clear progression of increasingly sophisticated tools that could be linked in time and place with *Homo* species that were becoming more and more modern-looking. But that is not the picture we have. Clearly, modern humans did evolve from earlier types in Africa but where and when the various stages emerged, and how they relate to the advances in stone-working is far from clear. Moreover, the most recent discoveries have raised more questions than they have answered.

It is one of the tenets of science, known as the law of parsimony or Ockham's razor, that we should not elaborate our explanations more than is necessary to meet the needs. On that basis we can confidently say that early hominins expanded their populations into new habitats whenever times were favourable because that is what all animals do, especially when adjacent territory is not effectively defended by rivals.

Early hominins are among many animals that originated in Africa and eventually extended their range into Europe and Asia. Hominins, sabre-toothed cats and the ancestors of spotted hyaenas may well have been fellow travellers.[15] The hippopotamus evolved in Africa and presumably followed other routes to become well established in places as far away as England. Hippo remains are so plentiful among the fossils found in East Mersea, Essex, that fossil beds dating from about 120,000 years ago, are known as 'hippo gravels'.[16] Hamadryas baboons present another interesting case, having evolved in Africa and colonized Arabia during the late Pleistocene (130,000–12,000 years ago). Analysis of their mitochondrial DNA indicates that there were population movements in both directions and that they took the southern route, across the Red Sea. It is possible that land bridges existed when sea levels were at their lowest.[17] So, while bigger brains and cultural advances were undoubtedly beneficial to early hominins, they are not necessarily key to understanding how and why they extended their range into Eurasia.

The first *erectus* fossils found outside of Africa were of 'Java man', discovered in the 1890s in Indonesia. These were followed by 'Peking man', found in the 1920s at Zhoukoudian, a productive site near Beijing that is now on the UNESCO World Heritage List. To some specialists, these and subsequent fossils indicate that Asian *erectus* types were sufficiently different from the African ones to justify renaming African *erectus* as a different species (*H. ergaster*).[18] Others lump them all together, accepting that *Homo erectus* shows a lot of variation.

Scientists have different opinions about how much variation to accept within an extinct species. Some are of a 'lumping' persuasion but others, referred to in the jargon as 'splitters', are more ready to regard consistent differences among fossils as evidence of more than one species. At this point we need a few words about the species concept.

Among living animals, species are groups of breeding populations that for some reason do not interbreed with other groups; they are reproductively isolated. When two or more populations have some degree of separation but will readily interbreed if the opportunity arises, scientists may consider them to be subspecies—not qualifying for full species status.

Speciation, that is to say the separation of one species into two or more different ones, can be easily understood as a consequence of geographical barriers. If two populations are kept apart, let's say by a river or mountain range, then different mutations can accumulate in the two populations to the point where the animals will not, or cannot, interbreed if they ever come together again.

Initially, a separation might be behavioural and affect courtship rituals or the mating season, or it could affect fertility or the fertility of offspring. Horses and donkeys are clearly different species with different chromosome numbers (horse, 64; donkey, 62) but they will mate and produce hybrid offspring that we know as mules when the father is the donkey. Male mules are always sterile and female mules are rarely fertile.

Among primates, hamadryas baboons and anubis baboons are quite different in appearance and they have different social structures but wild populations in Ethiopia hybridize where their home ranges meet. The offspring are fertile and a hybrid zone has emerged where the hybrids show a mix of the behaviours inherited from the parents.[19]

Obviously, drawing a firm line between species can be tricky, even with living animals, so it is little more than guesswork with types that have long been extinct. When there is no DNA to compare we have no way of knowing whether fossils of similar types were from populations that could interbreed or not. We shall return to this topic shortly but we must first visit another archaeological site.

Dmanisi is a little town in the republic of Georgia, south-west of Tbilisi, the capital. It has an ancient history as a settlement, and from its archaeological site there has been a revelation in prehistory. From among thousands of fossils of many different animals, a total of five hominin skulls were found between the years 1991 to 2005. The last of them was not fully described until 2013. They show brain sizes ranging from 600 to 775 cc and all five skulls are securely dated at about 1.8 million years. This means they pre-dated the Turkana boy by hundreds of thousands of years and also had smaller brains. Stone tools found at the Dmanisi site are all in the Olduwan tradition.[20]

The evidence is clear: hominins with a very low level of technology were among the animals that migrated and/or expanded their populations out of Africa more 1.8 million years ago. Were the hominins all one species, and if so, what should it be named? The team in Georgia, led by a distinguished anthropologist with the splendid Georgian name of David Lordkipanidze, can see no reason for calling it anything other than *Homo erectus*. The team would go even further and lump *H. habilis* in the same evolving group because there is no way to determine species within this highly variable lineage.[21]

So what happened next in human evolution? For three decades there have been two main schools of thought in answering that question. In the 'multiregional hypothesis' the populations that provided the fossils from Dmanisi were among those that spread throughout Europe and Asia and continued to evolve, together with any later arrivals, as a single species right up to the present day. The obvious racial differences that we see today have evolved in place but there has been constant genetic mixing to prevent any speciation.[22]

The alternative explanation can be called the 'recent African origin' model. In this scenario, populations such as those from Dmanisi did not give rise to the existing people of Europe and

Asia. Instead, continuing evolution in Africa produced anatomically modern forms like the ones from Morocco and the Omo River, mentioned at the beginning of this chapter. More recent fossils of 'early modern' types have also been found in South Africa. Populations of these relatively recent types spread outside of Africa within the last 200,000 years and it was from those that modern humans evolved.[23] In this scenario, it is a matter of speculation whether *sapiens* out-competed the *erectus* types in Eurasia, or whether *erectus* populations would have petered out anyway, failing to survive the climatic extremes.

More fossils have been discovered and new findings in genetics have been made since the two hypotheses were put forward. There is no shortage of literature on the subject and the evidence for both sides is raked over repeatedly with every new discovery that might be relevant. In this connection, molecular biology can now be more informative than fossils.

The last four species

During the past decade, mainly on the basis of molecular evidence, it has come to be accepted that no fewer than four species of humans have existed in Eurasia within the last 50,000 years. We have known about Neanderthals since the nineteenth century when the first fossils turned up in a limestone quarry in Germany. Then, on the strength of a fragment of finger bone found in Siberia in 2008, we learned of another group of extinct humans now known as the Denisovans from the name of the cave where it was found.[24]

A decade ago, part of a finger would have told us very little, but an impressive team of experts at the Max Planck Institute for Evolutionary Anthropology in Germany managed to recover DNA that had been remarkably well preserved in the cold of Siberia. This told us that the bone was from a little girl who died tens of thousands of years ago. A later analysis, published in 2012, revealed that her kind had less genetic variation than is found in modern humans—indicating that she had come from a small total population even by the standards of that time.[25]

The leader of the Max Planck team was Svante Pääbo, who was already world famous for his role in recovering the Neanderthal genome. The Neanderthals lived in large areas of Europe and western Asia but had disappeared by about 30,000 years ago. Over many thousands of years the Denisova cave was occupied successively by Denisovans, Neanderthals and modern humans so it was presumably regarded as prime real estate during Siberia's freezing winters.

Unfortunately, nobody has yet been able to salvage any DNA from the fourth species, *Homo floresiensis*, the bones of which were first found in 2003 inside a cave on the Indonesian island of Flores. The remains of a dozen individuals have been discovered in the same cave but that first find included much of an individual skeleton with most of the skull. It is believed to be that of an adult female who stood about one metre tall, had a brain volume of no more than 420 cc and lived as recently as 18,000 years ago. Inevitably, this species has become known as the 'hobbit'.

The hobbits cannot easily be dismissed as modern humans with a serious medical condition, though there are scientists who still believe that is a possibility. Taken together, the bones from different individuals show similar characteristics and they don't suggest pathology. From the balance of evidence to date, it appears that the hobbits were a tiny species, possibly related to the earliest types found at Dmanisi. Other mammals have evolved into dwarfs after isolation on islands, including pygmy elephants that lived on Flores at the same time as the hobbits. Research continues.[26,27]

As to the rivalry between the multiregional supporters and those convinced of a recent African replacement, all this uncertainty about species does nothing decisive. A recent African origin scenario can accommodate some interbreeding, and a multiregional explanation is still consistent with gradual assimilation of different types.[28] According to Svante Pääbo and his colleagues, Europeans and Asians have about 2.5 per cent of their DNA from Neanderthals, and only Melanesians show an additional few per cent of their DNA from Denisovans.[29] This geographical distribution raises still more questions but it does show that Neanderthals and Denisovans arose outside of Africa because there is no trace of them within that continent.

Toolkits and the end of an epoch

One might think that if we consider only the technology and culture of one species, we would at last find that advances were made more or less progressively, but again, the reality is not so simple. In Europe, the Middle East and Africa, tools of the Olduwan type had generally been replaced by Acheulean tools more than half a million years ago. But in China, and a few enclaves elsewhere, simple core and flake tools were still being used 100,000 years before the present and Australian core and scraper tools persisted until much more recently than that.

The Olduwan and Acheulean traditions constitute the lower palaeolithic, which was followed in Europe and the Middle East by a variety of stone flake industries representing the middle palaeolithic. Tools of the Neanderthals are a well-known example of this phase of tool making. The style is known as the Mousterian (after Le Moustier in France) and it wasn't entirely exclusive to the Neanderthals. To further complicate the story, equivalent flake industries, known by other names, occurred in Africa and much of Asia. The upper palaeolithic was the final technological phase of the early Stone Age, the transition usually taking a few thousand years. Flakes were generally replaced by longer blades produced from specially selected cores, although simple blades were not unknown in Africa and Europe at much earlier times.

In short, the named traditions in stone technology are associated with different phases in geographical areas rather than with the march of time across the globe. Moreover, even within a region, the records are often sporadic. For example, stone and bone tool traditions are well known from sites in South Africa beginning 75,000 and 65,000 years ago (Still Bay and Howieson's Poort respectively) but they each vanished after a few thousand years.[30] This lack of continuity is typical of the archaeological record before about 40,000 years ago.

Towards the end of the Pleistocene, in most regions, the toolkit of flakes and blades was being augmented with smaller specialized tools known as microliths that were used as spearheads, arrow tips or arranged in rows on composite tools. But that is not to say that they were appearing for the first time; some have long been known from sites in southern Africa that were occupied 40,000 years ago. Very recently, microliths were found in South Africa and dated at approximately 71,000 years.[31]

Metals finally brought an end to all stone tool industries but it did not happen within the Pleistocene and it did not happen quickly; stone tools were still in use in some parts of the world until recent times. And why not? The edge of a flake of certain types of stone, especially obsidian, can be extraordinarily sharp. Obsidian is a black volcanic glass that fractures to produce edges that are sharper than those of surgical steel scalpels.[32,33]

Obsidian must have been among the first items ever carried for trade because trade routes for chunks of obsidian existed 14,000 years ago.[34] Some might quibble about the definition of 'trade' in this connection, and we will come to that in chapter 6. Incidentally, if prostitution is considered to be a trade, then that must surely remain unrivalled in terms of antiquity, even though chimpanzees have not yet been seen to trade gifts for immediate sexual reward as is sometimes claimed.[35]

5 A hesitant transition

For eight thousand generations or more (generation time being an arguable quantity over that period) populations of anatomically modern humans survived as hunter-gatherers and remained viable enough, somewhere, to pass on the genetic code for our species. But it may have been a near thing. The people of Africa show more genetic diversity than do other people, indicating that after leaving that continent we passed through one or more genetic bottlenecks of low population size during which a lot of variation was lost. One bottleneck appears to have occurred when humans crossed the Bering land bridge from Asia into the Americas. Another, much earlier one, is indicated but there is no consensus as to when and why it happened.[1] Humans, of course, are not the only species to show evidence of population bottlenecks; cheetahs have so little genetic variation that they must have come close to extinction at some point.[2]

Pleistocene culture

By the end of the Pleistocene, people were well equipped to extract the most from the habitats that they occupied. They had an arsenal of stone cutters and scrapers, tips for spears, harpoons and arrows, wooden handles and shafts, and tools made from bone. There is no doubt that nets and crude traps would have been used, and fish could have been caught by damming streams; children still do it. No animals were too big to be hunted or driven to their deaths. Cooking had been a major breakthrough by *H. erectus*, extending the types of food that could be eaten and making all types of food more easily digestible. Now, seeds, nuts and other hard plant materials were being ground on stone slabs or with an early equivalent of pestle and mortar.[3]

Cave art, i.e. murals on cave walls, were first found in the nineteenth century in the cave of Altamira, in northern Spain. A description was published in 1880 but the images of bison, horses and other animals are so sophisticated that it was another twenty years before the experts of the time could accept them as being genuine palaeolithic art. The paintings, drawings, engravings and hand stencils were done over a period of some 20,000 years ending about 14,000 years ago. The cave is now a UNESCO World Heritage Site.

Many more cave paintings have since been discovered in Europe, and others are known from different parts of the world.[4] Naturally, they depict animals that occurred locally but there are also some common elements, notably prints and stencils of hands, and there is endless speculation about how it should all be interpreted. Beyond the evident artistic ability, what does cave art reveal about palaeolithic thinking, understanding and possible spirituality? It is a question that has occupied the entire working lives of scholars.

Ritual burials of the dead suggest there was a belief in (or at least a hope for) some sort of afterlife, and ritual burials were practiced throughout the period of cave art. Neanderthals also buried their dead with care—at least sometimes.[5] We can know nothing about musical talents from that time but flutes were certainly used, the oldest known being a flute with five finger-holes made from the hollow wing-bone of a griffon vulture. It was recovered from Hohle Fels cave in southern Germany and dated at about 40,000 years.[6]

The origins of cave art are still uncertain. A painting of a wild pig in a cave in Sulawesi, Indonesia, was recently found to be as old as the earliest large mammal paintings from Europe (about 35,000 years).[7] Other forms of decoration and artistic expression can be traced back much further in time, though of course opinions differ as to what constitutes 'art' no matter when it originated. *Homo erectus* evidently made zigzag scratchings on freshwater shells in Java around half a million years ago.[8] Finds in Israel and South Africa, 100,000 years old, reveal that beads were being made from shells, and ochre was being ground to a paste for paint at about that time.[9] Shells evidently had a special significance that was widespread across cultures for tens of thousands of years. Shells from the Mediterranean and Aegean seas came to feature prominently in trade or exchange networks that extended for hundreds and even thousands of kilometres from their source.[10,11]

Personal ornaments and other portable works of art appear to have been popular during the time of cave art. These include clay models and decorative carvings in bone, antlers and ivory. More than a hundred of the much-publicized Venus figurines have been found in Europe, all obviously female with exaggerated female shapes but usually lacking arms and feet. The heads are generally small and faceless. Did these little figures have some special, symbolic meaning or were they made just for the fun of it? We can never really know.

What is certain is that during the period from about 40,000 years ago to the end of the Pleistocene (roughly the upper palaeolithic of Europe and the Middle East) there was relatively consistent cultural advance compared with the patchy and sporadic record of human activities from earlier times, when ice or drought may have effectively depopulated areas for thousands of years at a time. Populations at the end of the Pleistocene were not being subjected to such lengthy and extreme climatic disruptions and by that time advances in skills, equipment and social organization may well have enabled families and bands to survive hardships that would have been fatal to earlier groups.

The extent to which people caused the disappearance of large mammals of that period is yet another topic that sparks arguments. The idea has been controversial since it was put forward by Paul Martin in 1973.[12] It is likely that some species were seriously over-hunted if not completely exterminated; the timings of human migrations match up suspiciously well with the disappearance of so-called 'megafauna'.[13] Species that had never encountered humans might have been pathetically easy to kill whereas African species that had co-evolved with humans would be another matter.

The extermination of valued prey animals would be a loss to the hunters but the losses could be sustained for as long as there were alternatives to fill the dietary gaps. Not even the most efficient hunter-gatherers could be sustaining their own populations if they were seriously diminishing their overall food supplies at the same time. For all their technical advances and capabilities, people throughout the Pleistocene were still entirely dependent on the productivity of the natural ecosystems of which they were a part.

Into the Holocene

The last glacial maximum (26,500 to 19,000 years ago) was followed by several thousand years of interglacial conditions when ice sheets were melting, sea levels were rising and the world was warming up. But during this interglacial there was a blip, called the Younger Dryas, that has puzzled scientists for some time. The name 'Dryas', incidentally, comes from the generic name of a plant (*Dryas octopetala*) that became common in Europe during the period. The Younger Dryas was the time between about 13,000 and 11,7000 years ago during which the northern hemisphere returned to near glacial temperatures. The most recent explanation is that meltwater was released, like a dam burst, from an enormous glacial lake in North America. On entering the North Atlantic this slowed the thermohaline circulation—the deep ocean conveyor described in chapter 3.

The Younger Dryas ended abruptly 11,700 years ago, when Greenland temperatures rose by as much as 10° Celsius in only a decade.[14] In the northern hemisphere this marks the end of the Pleistocene and the beginning of the Holocene—the epoch in which we now live. After that, the interglacial resumed its course and the Sahara Desert became green and productive as it had presumably done many times before when the precession cycle presented the northern hemisphere to the summer sun.

The start of our epoch, as a point on the geological time-scale, was not a time of very sudden change in human affairs. Rather, it was a time of opportunity for human populations so the patchy and faltering pattern of social change may seem surprising. This is not to suggest that the Holocene, anywhere, has been twelve thousand years of stable and ideal conditions but substantial parts of the world did became richly productive once the ice had retreated. Yet the human way of life was not swiftly transformed. From an ecological point of view, this makes the transition to agriculture all the more interesting. It is not easy to generalize but specialists are agreed that the early Holocene was generally warmer than the twentieth century, with the southern regions apparently reaching their maximum warming a few thousand years ahead of the northern hemisphere. Long-term climate changes within the epoch are consistent with the effects of the Milankovitch cycles, modified by the vast North American ice sheet which did not disappear until around six thousand years ago. There were significant short-term variations of smaller amplitude than had occurred during the Pleistocene, but overall, the Holocene appears to have been by far the longest period of warmth and stability in the last 400,000 years.[15] Initially, human populations became more widely distributed as more land became habitable, and densities would have increased as occupied regions became more productive, but after a few generations, as populations approached the higher carrying capacities, growth would slow for the reasons described in chapter 2. The entire human population in the early Holocene probably averaged no more than about seven million—which is smaller than that of present day London or New York.[16]

In the short term, food supplies can be profoundly affected by nothing more drastic than a period of unfavourable weather, so families and bands of hunter-gatherers could be struggling to meet their needs after two or three particularly poor seasons. This is a familiar situation to many animals. In lean times they must adjust their diets and work a lot harder to survive. They know what they can eat, if they must. There are birds in my garden that will squabble during droughts for food that they would ignore at other times. It is a simple demonstration of supply and demand.

Wild boars are as omnivorous as humans, even though they can't catch grown birds or go fishing. Like hunter-gatherers, their diet varies according to geography, season and degree of competition for food but they show their preferences when they have the opportunity. A study at two sites in the Netherlands found that tree seeds (notably acorns) made up the main autumn and winter subsistence but, failing that, the boars of one site survived mainly on broadleaved grasses, and in the other site, on broadleaved grasses, hairgrass and roots.[17] Roots and bulbs are a major part of the boars' diet in the French Alps.[18] In general, more than 90 per cent of the diet is vegetable but wild boars do seem to need some meat, and during their grubbing they find a variety of creatures, such as earthworms and other small invertebrates. On frozen ground or in deep snow in northern Europe they obtain meat by scavenging at carcasses of other mammals.[19] On the beaches of Queensland feral pigs dig up and eat the eggs of turtles. In summary, wild boar and their descendants are highly adaptable and are great opportunists.

Humans in the early Holocene would also have been adept at extracting all their needs from the natural environment. In familiar territory they would have had an intimate knowledge of what was edible, where and when it could be found and how it could be collected or killed. In addition, they had the technology to cook, dehusk, detoxify, grind and otherwise process plant materials as they chose. Above all, they had a uniquely human social structure that included 'property rights' to extend, when necessary, a measure of cultural control over the killing and collecting.[20]

That is not to say there is anything especially human about property rights that are maintained by a group having the ability to defend core territory within a home range; defence of territory emerged long before there were people to think about it. But it is more important among those species that make investments in their territory because then there is more at stake. Some species spend a lot of time and energy in building elaborate homes or nests and we know that humans in the early Holocene practiced a form of proto agriculture in attempts to make ecological conditions more favourable for themselves. Vegetation was strategically burned, and particular food plants may have been singled out for some sort of special treatment such as spreading bits of tubers around a parent plant. There is no point in doing this if the benefits will be reaped by competitors roaming over common ground. Aspects of the problem that Garrett Hardin described as the 'tragedy of the commons' in 1968 have very ancient roots, though the tragedy is now more familiar in the context of over-fishing the high seas and polluting the atmosphere.[21]

When times were good the people of the early Holocene could eat what they preferred with the least effort. At other times, and increasingly so as populations approached carrying capacity, they would need to survive on less favoured food and possibly spend a lot more time obtaining and processing it. Spending more energy for each unit of energy gained is known to archaeologists and anthropologists as 'subsistence intensification'. And in the literature, for nearly fifty years, much has been made of the discovery that the transition to farming in south-west Asia involved a period of subsistence intensification that led to a much broader range of foodstuffs being included in the diet. This, it is said, represents a 'broad spectrum revolution' in palaeolithic diets.[22] In the Mediterranean region, for example, and at sites in Israel, hoofed animals were by far the main source of meat for hunters of the middle palaeolithic, but in the upper palaeolithic the prey included (though not invariably) many more small animals and birds that were harder to catch for much less reward.[23,24,25]

Similarly, a site on the shore of the Sea of Galilee, in Israel, has revealed that in the last glacial maximum, about 23,000 years ago in the upper palaeolithic, the occupants were including at least 142 types of plant in their diets. Fruits and seeds such as legumes and acorns were favoured but many of the plants were grasses; not only wild wheat and barley but also a large proportion of species with very tiny seeds. In later pre-agricultural sites, with a much higher proportion of wild cereal grasses, there were far fewer small-grained grasses.[26] As to the mix of grasses found in sites from the middle palaeolithic, the evidence is inconclusive but there is at least the indication of broad spectrum exploitation of plants by Neanderthals more than 55,000 years ago.[27]

From an ecological point of view there is nothing surprising about these findings; they demonstrate the familiar response of omnivorous animals whose populations are sometimes, but not always, close to k because of density and/or environmental change. I can only say there does not appear to be any sort of dietary 'revolution' involved.

Why grow your own food?

The earliest and best-known transitions from hunting and gathering to farming occurred in the region known as the 'Fertile Crescent'—a term coined by an archaeologist in the early 20th century. The area between the Tigris and Euphrates (known to the ancient Greeks as 'Mesopotamia', between the rivers) forms the eastern side of the crescent, which arches westwards in the north, through the Levant and the eastern Mediterranean, and takes in the Nile Valley and Nile Delta as the western arm of the crescent. Today, depending on how the boundaries are drawn, about a dozen nations are at least partly included in the region.

Numerous archaeological sites have provided snapshots of life at different times in and around the camp-sites and settlements and there is no shortage of opinions as to how they should all be interpreted. But, as usual in the evolution and prehistory of our species, there is no steady progression, nor is there a clear pattern of events that is common to all centres around the world where agriculture was initiated. In this context, there is not even an agreed definition of agriculture—some authors emphasize the social and labour requirements involved, others focus on the domestication of crops and livestock, and some see the degree of dependence as the important criterion. But to begin with, it must always have been subsistence agriculture—by which families were simply trying to feed themselves.

Without doubt, both physical and mental abilities had to be adequate to sustain lengthy projects, and in each case there had to be available wild plant or animal species that were suitable for domestication. In the case of plants, the favoured species would need to be cultivated for some time to produce the domesticated strains that have appeared in the archaeological record. But once people had begun collecting and sowing home-grown seed there would be automatic selection for genotypes that did best under those circumstances, so to begin with, selection for new strains need not have been a conscious effort.

The most contentious issue, however, is motivation. What was the incentive for communities to start struggling with a new way of life and to persist with that struggle? It seems most unlikely that the first generations of farmers, sowing wild seeds, using only the simplest of tools, and with no traditions to guide them, would have found much encouragement in the quality or quantity of their harvests.

Before going any further we need to clarify a few terms that refer to the broad timing of cultural change at this stage. Since the nineteenth century it has been usual for non-specialists to divide the Stone Age into three phases: the palaeolithic, mesolithic and neolithic. In this simple scheme the neolithic is characterised by polished microliths, specialised stone tools and early agriculture, with or without pottery. The palaeolithic, on the basis of stone technology, is subdivided into lower, middle, and upper palaeolithic but we have seen how these stages overlapped and varied in time around the world.

It is now generally accepted that there is no clear distinction between any of the stages, even within one region. In the Levant the mesolithic and neolithic have been described as a continuum 'demonstrating a temporal mosaic of cultural change that extended for more than 10,000 years'.[28] The term 'epipalaeolithic' (instead of mesolithic) was introduced early in the twentieth century but came to be used differently by experts working in different parts of the world so did nothing to reduce confusion for non-specialists.[29]

Similarly, the term 'Natufian' was adopted early in the twentieth century to refer to socially-organized communities of settled hunter-gatherers in the Fertile Crescent. Natufian cultures are recognized from the late Pleistocene until well into the Holocene, which is a period spanning about 2,600 years between 12,800 to 10,200 years ago. The name comes from Wadi en-Natuf, a watercourse north-west of Jerusalem. The core homeland of the early Natufians was the eastern Mediterranean, where the vegetation was a mosaic of open forest and grassland. Their homes were typically rounded pit houses with floors dug below ground level and surfaced with stone. They were presumably roofed over with branches and simple thatching. Sites consisting of clusters of these huts might have been inhabited continually for long periods but would easily be restored if they were used as seasonal base camps. It was productive country with good supplies of wild cereals, acorns, pistachio nuts and a variety of animals, especially gazelles. Yet it has been estimated that bands of hunter-gatherers in that time and place would have foraged and hunted over areas of 300 to 500 square kilometres.[30]

Carrying capacity was drastically reduced by the Younger Dryas so many Natufian sites were abandoned. Some communities became more nomadic and moved away to the south and west. Those that remained evidently managed to adapt and survive. Their descendants are believed to have been among those whose settlements included cultivated fields when, centuries later, the Younger Dryas relaxed its grip. The Israeli archaeologist, Ofer Bar-Yosef, spent decades researching the subject, and as he puts it, 'The beginning of intentional widespread cultivation was the only solution for a population for whom cereals had become a staple food.'[31]

However, as Bar-Yosef also points out, 'no field project outside of the Levant has yet exposed any indication of a prehistoric entity that resembles the Natufian.'[32] So, interesting as they are, the Natufians don't provide a model for explaining the origins of agriculture elsewhere and the term 'Natufian' may have no wider, useful application.

One might at least feel confident in saying that settlement must have come before agriculture but it seems not always to have been the case. Nor, as was widely believed, can we say that cultivation invariably came before herding livestock. Among the lakes and streams of the green Sahara, in the early Holocene, it appears that cattle herders were at least seasonally nomadic and were hunting and collecting wild grass seeds, not cultivating crops, some 9,000 years ago.[33] And while settlements, crops and livestock may have came before pottery in the Levant, there were mobile hunter-gatherers carrying pots in other regions, including the Sahara, south Asia, and among forager-fisher communities in South America.[34] The Jomon people of Japan were exceptional in making pottery for thousands of years before there was any evidence of cultivation.[35]

Agriculture is believed to have started independently in no more than nine centres around the world: the Fertile Crescent, north and south China, Mexico, the Andes, eastern USA, and sub-Saharan Africa. Some authorities also include New Guinea and Amazonia. In each case a few local plants became the founder crops. In the Levant, these included wheat and barley, lentils, peas, chick peas, an oilseed and the flax plant for fibre.[36] Sheep and goats appear to have been the first domestic animals, with cattle and pigs following. Maize and beans were among the main crops in Mexico, whereas rice was a founder crop in China, and our potatoes came originally from the Andes.[37]

From these centres, the knowledge, skills and crops, spread throughout the world along exchange networks and by the migration of people—which also transferred human genes and influenced languages.[38] But returning to the question of why people started growing crops and domesticating animals in the first place, no single, inclusive answer has been found. There were many ecological similarities, for example, between parts of the Fertile Crescent, where agriculture evolved early and relatively swiftly, and the western part of North America—where hunting and gathering persisted until European settlement.[39]

After reviewing a dozen or so different approaches to the problem, the American anthropologist, Alan Simmons, sums up as follows: 'Is there, then, any consensus of the characteristics associated with the emergence of agriculture? The simple answer is "no," although some scholars feel that there is a core of recurring traits.'[40]

It seems safe to assume that any individuals attempting to cultivate crops would need to be supported for a time by hunting and gathering; they could not have abandoned it abruptly. Consequently, farming may have been an on-and-off sort of enterprise for very long periods, with neighbouring communities sharing their experiences for generations. In parts of America the two ways of life ran concurrently for thousands of years, as they still do in the forests of West Africa and South America, but we'll come to that later.

We can also say that communities outside the initial centres were not always keen to try the new ways when the opportunities arose; people in northern Europe, it seems, were particularly reluctant.[41] Some native Americans adopted agriculture but then abandoned it and reverted to hunting bison. Evidently, their numbers could still be sustained in that way and they could do it on horseback after horses were reintroduced to America by the Spanish.[42]

And this brings us to a most important point: when people were supported by agriculture, they could continue to increase their populations while they impoverished or even destroyed the natural ecosystems of their homelands. Never before had this been a possibility. As a consequence, as the most productive lands were converted to agriculture or used for herding livestock, human populations increased, carrying capacities in the adjacent wild were often lowered, and the option for communities to revert to hunting and gathering was progressively lost.

Not only did farmers and pastoralists close off their own options in this way, but over time, adjacent hunter-gatherers would also be obliged to grow and/or herd their own food whether they liked it or not. As areas became more densely-settled, hunter-gatherers were soon locally outnumbered, and as Peter Richerson and others have pointed out, they would be quite unable to compete for survival by hunting and gathering while wild resources dwindled and they continued to lose land to settlers.[43]

For whatever reasons to begin with, over a period of a few thousand years most people would come to depend mainly on the continued success of some sort of agriculture. For them, the wilderness was no longer home and the course to the future was firmly set for a point of no return.

6 Farmers, towns and temples

The inherited characteristics of our species, from sweat glands to sperm counts, brain modules to bowel movements, became part of our genetic make-up because they favoured our hunting and gathering existence. Evolution could not have prepared us for farming any more than it has prepared us for air travel or desk jobs. Only during the last few thousand years—a fraction of one per cent of human evolutionary time—has anybody grown up learning how to live in a different way.

Some characteristics that served us well in the past are now a lot less useful. For example, infants effortlessly learn the sounds of speech during a sensitive period that begins in the womb and ends at about one year, though the sensitive period for syntax runs through the first few years.[1] That was ample time, no doubt, when the world was as big as your homeland but it's not ideal for mature students who find they need a foreign language. Similarly, with a high chance of being wounded during a short but healthy life, it was best to form blood clots quite swiftly. But this increases the risk of strokes and heart attacks in old age so many of us now need a daily dose of anticoagulant to improve the prospects for another decade. One of the genes involved in blood clot formation, incidentally, was inherited from Neanderthals so is a consequence of interbreeding.[2]

Whatever the pros and cons of our genetic make-up, the fact is that those who walked on the Moon carried the same ancient genome as those who slaughtered mammoths. The first farmers could not have dreamed of the changes that would become possible as a result of their new way of life, and judging by the archaeological evidence, the immediate challenges of providing food and shelter must have seemed more than enough to cope with at the time.

Snapshots from the neolithic

As with the transition from foraging to farming, when a broad sequence of events is outlined in a few paragraphs it can suggest a progression towards a goal, a journey towards a destination. In reality, neolithic settlements, ranging from scattered clusters of semi-permanent huts through to sizeable villages, existed at the same time in different places. There is nothing to suggest a steady progression towards organized townships, although the sort of social organization that existed in each place can only be inferred from the material evidence that comes to light.

In this context, the settlement of Tell Abu Hureyra, in northern Syria, has been intensively studied because it was continuously occupied for thousands of years. The word 'tell' in the name comes from Arabic and refers to the mound that builds up over millennia when successive settlements are built over the ruins of earlier ones. The tell of Abu Hureyra covers an area of 11.5 hectares to a height of about eight metres. Many ancient settlements in the Middle East became raised on tells.

Abu Hureyra was first settled around 11,500 years ago when a few pit houses were dug into the soft sandstone. For about 1,500 years (the period known to archaeologists as Abu Hureyra 1) the site was continuously occupied by small numbers of hunter-gatherers, but then, probably because of drying effects of the Younger Dryas, it was largely abandoned. Some three centuries later it began to be used intermittently, and by 9,000 BP (before present[3]) Abu Hureyra 2 had grown to be a permanent agricultural settlement where thousands of people lived in rectangular mud-brick houses.

Abu Hureyra 2 flourished as a large village for another two thousand years. Throughout that period the people cultivated cereals and pulses but a change in the source of meat occurred quite abruptly. Until that point most meat still came from the wild. Goitered gazelles (*Gazella subgutturosa*) were evidently killed and butchered *en masse* after they were driven into long, stone-walled traps built along their migration routes.[4] But then, around 8,300 BP, there was a switch to the herding of sheep and goats.

Excavations of Abu Hureyra took place in 1972 and 1973, after which the site was flooded by construction of the Tabqa dam on the Euphrates River. Accurate dating of stored artefacts was completed during the 1980s when the British Museum and Oxford University collaborated in using a technique (accelerator mass spectrometry) by which radiocarbon dating can be done on the tiniest amounts of material.[5,6]

Note that Abu Hureyra is never described as anything more than a big village, not a town. So when does a big village become a small town, or a big town become a city? It is a question that raises many other questions about social organization and hierarchies of power and influence within and among communities—topics about which an immense amount has been written. In his four-volume treatise on *The Sources of Social Power*, sociologist Michael Mann begins by discussing societies as organized power networks and then points out that the concepts he develops under that heading have 'virtually no relevance to 99 percent of humanity's life span to date'.[7] He has a point. What is there to say about organized hierarchies of power among hunter-gatherers? As far as we know they knew nothing of elites or classes beyond their immediate pecking order.

In settlements where there was some organized power structure, but before writing was invented, we have only the evidence of buildings and other material remains to tell us how society worked. Gordon Childe, one of the most influential archaeologists of the last century, pointed out that the concept of 'city' is notoriously hard to define but he went on to list ten criteria that would 'serve to distinguish even the earliest cities from any older or contemporary village'. In other words, while we cannot define the first cities in terms of any particular detail, we can recognize them as part of the broader transformation that Childe, a passionate Marxist, chose to call 'the urban revolution'.[8] The word 'revolution' seems to have broad appeal and, arguably, this is another case of its misuse. But whatever we call it, the rise of the state was mainly based on new social institutions and practices that differed in particulars but served the same ends. Perhaps, without losing too much of the sense of it all, I can list the criteria as ten main points:

1. Population size and density would have been greater than previously.
2. Some inhabitants would not be food producers; they would be, for example, merchants, craft workers, officials and priests.
3. Primary producers would be obliged to provide surplus as tax or tithe to an authority who would accumulate it.
4. Monumental public buildings would distinguish each city and symbolize the social surplus.
5. The accumulated surplus would be used to support those not directly involved in food production.
6. It was necessary to invent systems of recording, if only for administrative purposes.
7. Calendars and mathematical achievements are features of the earliest civilizations, and are a corollary of writing.
8. Artists of all kinds developed conceptualized and sophisticated styles specific to different urban centres.

9. Foreign trade expanded. Compared with earlier villages, first cities were far more dependent on essentials from afar.
10. Specialist craftsmen were provided with raw materials and could be secure residents, not itinerants. But they were dependent on temple or court and had lower-class status.

Even if we accept this check-list without quibbling there is still room for argument about which settlement should qualify as the world's first city if indeed there is a clear first. Jericho, in what is now called the West Bank, is probably the best-known contender, the first inhabitants having settled about 11,000 years ago, though Jericho has not been continuously occupied since then. I shall leave it to others to argue about world record-holders; I shall use just a few examples as stepping-stones through the remainder of prehistory.

Çatalhöyük, near the town of Konya in present-day Turkey, is an exceptional neolithic site of 13 hectares that at one time might have housed as many as 5,000–8,000 people. Two tells rise as much as 20 metres above the plain and reveal up to eighteen layers of settlements from times between about 9,400 and 8,000 years ago. The small, (average about $20\,\mathrm{m}^2$) mud-brick houses are roughly square or rectangular and some of them are subdivided into one main room with the hearth or oven, and one or more tiny storage rooms. There are thought to have been somewhere between 1,000 and 1,600 houses when Çatalhöyük was at its biggest.[9,10]

In the earlier levels, at least, houses were built together in tight clusters with no street space between them so access to each house in a group must have been through the roof. Alleys or middens between clusters appears to have separated numbers of these tight-knit communities from each other. Those sharing common rooftops were presumably on very close terms so are likely to have remained within the numbers that allow such relationships to be comfortably managed (chapter 2).

In the wall-paintings and other artefacts, such as skulls and horns or claws of wild animals, there is ample evidence of symbolism that extends to the treatment of the dead. About one in six of the buildings was used for multiple burials so it seems that the families of each cluster used certain long-established houses, now termed 'history houses', to bury their dead. But there was no clear division between the homes of the living and buildings that served as cemeteries; skulls and sculptures might be buried together for a time, then dug up and later reburied. It appears that 'the houses at Çatalhöyük and all the activities that took place in them were seamlessly religious, social and practical. The lived-in houses were religious in the sense that they were about imagining, remembering and interacting with past houses and those who lived in them.'[11]

Çatalhöyük, despite its size and levels of social integration, was never more urbanized than was Abu Hureyra in the sense that there were no public buildings or hierarchies of power and influence. Society was divided only by degrees of familiarity into families, close communities and neighbourhoods. The site is listed as World Heritage by UNESCO and is still being excavated as well as being open to the public.[12] The Turkish name, pronounced *cha-tahl-hew-yewk*, translates as 'forked mound' and refers to the double tell.

Çatalhöyük fell into decline and was abandoned about 8,000 years ago, so if we want to see the full transition to urbanization we must look elsewhere, and nowhere is better documented than Mesopotamia—the land between the rivers.

Mesopotamia

This area today includes Iraq and parts of Syria and Iran. Both the Tigris and Euphrates rivers originate in Turkey and flow south-eastwards to the Persian Gulf. Great flood plains exist between the rivers in the south, where thick beds of fertile silt are alternately flooded and sun-baked for months on end. The plains were highly productive agricultural land for settlers who had boats and knew how to irrigate crops. The two rivers eventually join as a single waterway—the Shatt al-Arab—that winds through the marshes and now forms the political boundary between Iran and Iraq.

On the plains of southern Mesopotamia, in 1919, an archaeologist from the British Museum discovered a distinctive type of pottery at a small site called Tell al-Ubaid. Accordingly, the style of pottery became known as the 'Ubaid' The pots had zigzag or wavy black lines painted on a buff-coloured background.

In the decades that followed it became apparent that Ubaid pottery, or something very similar, could be found throughout the whole of southern Mesopotamia. Modern radiocarbon dating suggests that, by some definitions and in some regions, the Ubaid spans a period of nearly 3,000 years—from about 8,500 to 5,800 BP.[13] For shorter periods, and with regional variations, the same sort of pottery extended over a vast area, including the whole of Mesopotamia, south-eastern Turkey, northern Syria and western Iran. The whole region was linked by trade routes, not only along waterways but over all terrain accessible to pack animals. Caravans of donkeys travelled hundreds of miles. And, of course, styles are dispersed and modified by copying so quite possibly, instead of having a single source, 'the elements of an Ubaid identity emerged from a vast inter-regional melting pot of influences.'[14]

Archaeological periods are commonly identified by their characteristic artifacts, especially stone tools and pottery styles, but the term 'Ubaid' now has wider meaning. It is a word that we find in the literature, again and again, not so much with reference to pottery and particular time-frames but to a whole regional period of social and cultural change. It was during the Ubaid that settlements really began to approach Childe's criteria for a city. Most people still lived in villages and small communities but towns began to grow and include bigger houses and there were obvious temples with more elaborate architecture. Many towns had acquired defensive walls by the end of the Ubaid.

Scholars disagree about how much authority was centred on the temples, and about the extent to which townsfolk were still more or less equal in status. Possibly there were differences between north and south in this connection but in any case it was only a matter of time because, throughout the Ubaid, the original egalitarian way of life was inexorably, if erratically, coming to an end.

The pottery style that replaced the Ubaid did not appear as any sudden innovation. It was more a case of the Ubaid style of domestically-crafted pots giving way to a more casual, unpainted, sort of pottery that was becoming mass produced—if not actually disposable. The change in pottery was the main indicator of the transition from the Ubaid to what archaeologists call the Uruk period. But before going any further in that direction we need to clarify a few other terms.

It is unlikely that the people of the Ubaid had any collective name for themselves or for the broad region that they came to occupy. The many different groups probably distinguished themselves with separate names—possibly with reference to their own particular homelands. The Greek word, 'Mesopotamia', did not then exist. In the earliest written references, the southern part of Mesopotamia was known as 'Sumer' to the people further north (the Akkadians) so they called its people Sumerians—though it is not the name the southerners used for themselves. The Sumerians, according to modern scholars, called themselves by a name meaning 'black-headed people' in their own language.[15]

And it is their language that sets the Sumerians apart. We know this from written records because they invented a form of writing that would last for nearly three millennia and would almost certainly give rise to Egyptian hieroglyphics. I call it a form of writing because it is not a language or even an alphabet; it is a system that uses between 600 and 1,000 characters to spell out words and sounds so it can be used for modern languages just as well as Sumerian. Scribes did the writing by pressing the tip of a reed into wet clay. The surface of any clay object would serve but most writing was done on palm-sized tablets. The tip of the reed was cut to a wedge shape, which explains the term 'cuneiform writing' (Latin, *cuneus*, wedge). All urban centres had their scribes but most other people remained illiterate throughout the history of cuneiform.[16] The British Museum holds a huge collection of original tablets and has recently produced an illustrated paperback on cuneiform.[17] It looks appallingly difficult to learn.

So the Sumerians were a distinct linguistic group but were they indigenous to southern Mesopotamia or did they move in from elsewhere? Nobody knows and, for those who care, it remains the most troubling Sumerian question.[18] What is known is that they were a people who moved rapidly (relatively speaking) through urbanization to the city state, and because they left plentiful written records their progress is far better documented than that of their predecessors. That is not to say there were no written records before the Sumerians; an archaic, or proto-cuneiform script, was used earlier during the Ubaid but there are far fewer examples and it is still poorly understood.[19]

We can now return to the period, and pottery style, known as the Uruk—which is the Akkadian name of one of the main centres of Sumerian urbanization. In modern Iraq, the ancient site has the Arabic name *Warka*. From its beginnings, roughly 6,000 years ago, Uruk expanded over a period of some 800 years to be a city inhabited by an estimated 50,000–80,000 people.

Southern Mesopotamia had tens of thousands of square kilometres of irrigated plains with towns, villages, hamlets and trading outposts connected by trading routes and waterways. There were other Mesopotamian cities contemporary with Uruk but scholars are generally agreed that Uruk was pre-eminent among them and was the main centre of trade and influence. Exactly how the influence was maintained is far from being generally agreed, but when some sections of society are working full time in occupations that produce no food, then they must obviously be fed from the surplus provided by others who toil in the fields. Regulation of this food supply in cities is fundamental to urban life, whichever way it is maintained. One way or another, economy comes to mediate human ecology so from early times the economic value of the farmer's harvest has been important in ways beyond the nutritional needs of the farming family.

In the view of Petr Charvát, a Czech archaeologist, Uruk was admirably democratic in this regard. It was a world in which:

> The material standard of living is equalized by means of redistribution, external threats are eliminated by garrisons posted to ward off any attack. Everyone is close to everyone else, people meet in assemblies to discuss and decide matters of common interest with at least some of the resolutions put down in writing. All receive the same treatment, both in life and death.[20]

Other researchers offer very different interpretations, mostly accepting some measure of administrative and religious control by elites, possibly extending to a sort of colonial control over the region centred on Uruk. Susan Pollock reflects on the extent to which hunger was a result of deliberate strategies to extract tribute. In her view, 'Quite likely hunger was an ever-present threat that also became a way to discipline an emerging workforce whose members no longer controlled either their own labor or the food they ate.'[21] So scholars could scarcely disagree more about the power structures that prevailed during the Uruk, but again, it was only a matter of time. The exercise of power by elites would eventually become conspicuous and indisputable.

The ruins of Uruk cover five and a half square kilometres with remains of a massive perimeter wall that must have been at least nine kilometres long and is reported to have been over seven metres high and studded with gates and towers.[22] There is clear evidence of frequent building and rebuilding throughout the site. The earliest layers show the familiar painted pots and high-fired clay sickles of the Ubaid but they are absent from later layers when the Uruk culture is characterized by crude, mass-produced clay containers and monumental buildings that we call 'temples' for want of a better word. We must bear in mind that by this time there was a pantheon of gods and a rich lore of myths and legends. Temples existed in many centres in Mesopotamia and Uruk had a number of them.

Exactly how the early temples were used is another contentious subject. They appear always to have served as combined religious, ceremonial and administrative centres but economists have put forward very different views about the degree of central control and the significance of free trade—topics to which I will return in chapter 7. The Uruk site is centred on the temple of Eanna, home of the goddess Innana. Gwendolyn Leick, in her popular book on Mesopotamia, describes the architecture of Uruk as 'stunning'. The temple itself was built of limestone blocks and was 62 metres long and over 11 metres wide.[23]

But, impressive as the buildings must have been, it is the artefacts within them that tell us most about life at the time. Bureaucracy burgeoned during the Uruk, which must have made huge demands on the prevailing writing and official stamping technologies. Stamp seals were used in the Ubaid but cylindrical seals were invented in Uruk. They were not much bigger in the hand but they applied a larger imprint when they were rolled across a surface of wet clay.

Designs on the seals include recognizable animals, both wild and domestic, and people who were engaged in a variety of activities—including 'men armed with spears, bows and arrows, or maces, guarding bound prisoners'.[24] Much of what is known has been gleaned from images on the seals together with the information on hundreds of thousands of clay tablets. To begin with, the scribes seem to have been entirely occupied with accounting, administrative work and tutoring trainees, but as the system improved it was widely applied to royal inscriptions and personal correspondence.

A much-quoted Standard Professions List names about a hundred different occupations including priests, courtiers, ambassadors, supervisors, smiths, jewellers, stonecutters, carpenters, weavers, bakers, farmers, herders, fishermen and woodcutters.[25] Ubiquitous and particularly crude clay bowls, referred to by today's writers as 'bevelled rim bowls', appear to have been made in three sizes (all under 1 litre) and may have been used for providing workers with food and/or beer.[26,27] This is not to imply, however, that all workers were employed institutionally.

Nor does the utilitarian pottery indicate a lack of skill; it merely shows that time and skill were not to be wasted. Sumerian craftsmanship was impressive, as was the range of materials that could be worked. No doubt local reeds and clay were used in every way possible but luxury metals and stones, as well as some essentials, had to be brought from far beyond Mesopotamia in exchange for textiles and agricultural products from the southern alluvial towns.[28]

A huge amount of timber would have been needed together with stone and bitumen for construction. Less common stone, such as quartz, marble and obsidian, was used for statues, containers and smaller items. Weapons and implements were made from copper, a versatile metal that was eventually alloyed with tin to make bronze, though the first bronze was probably a mix of copper and arsenic that more commonly occurs naturally.[29,30] Considering the transport and labour involved with the prevailing level of technology, all this production and trade must have kept many people, including scribes, working for very long hours.

The influence of Uruk began to wane as other urban centres increased in size and power to become independent city-states ruled by hereditary dynasties. There was still economic exchange throughout the region but during this Early Dynastic period (roughly 4,850–4,300 BP) the city-states of southern Mesopotamia were rarely peaceful, and were sometimes at war.[31] With the hindsight of history, we should not be surprised that Uruk and other city-states of that time lost their independence completely as soon as they could be conquered and unified by a more powerful force. Peace among men, it seems, has never been the default state.

Sargon, King of Akkad ruled the region (and far beyond) for 56 years, according to the Sumerian King-list, and so Uruk became part of the Akkadian empire for a time. Thereafter its fortunes waxed and waned but it was not entirely abandoned until sometime during or after the Muslim conquest of Mesopotamia, less than 1,400 years ago. In this chapter we need go no further; urbanization was firmly established, the Bronze Age had arrived and the first civilizations, as we know them, had emerged.

7 Explaining urban expansion

In chapter 5 we saw the difficulty of explaining why some communities struggled for generations to succeed with agriculture when their predecessors had been well-fed and adequately sheltered by working for just a few hours a day. Urbanization presents a similar puzzle. Today we can all see the benefits that city life can bring but thousands of years ago nobody could possibly have imagined the advantages of living in a modern city.

So why did people take to crowding together in bigger settlements and accepting, in most cases, a loss of independence and the apparently gruelling working lives that went with it? It is easy to see that at higher densities of population, made possible by agriculture, people would be locked into a settled, farming existence, but why continue with urbanization to the point of creating city states? Was it simply a matter of security and finding strength in numbers? If not, what were the immediate rewards? A person's well-being, then as now, depended largely on nutrition and health so we'll look first at those.

Diet and nutrition

Over a period of some 3,000 years, the Natufians, living at low population densities and eating a wide variety of plants with lean meat mainly from gazelles, appear to have been a relatively healthy lot. In the same region, the earliest farming communities of the neolithic continued to hunt and collect wild foods for a long time before they became fully dependent on agriculture, and there were no sudden changes in health or stature judging by analyses of skeletal remains.[1]

An adequate diet includes enough food to keep us fuelled by ATP, and a supply of all the vitamins and minerals that are essential for normal operation of human cellular machinery. Of the three classes of food (carbohydrates, proteins and fats/oils) the fats/oils, collectively called lipids, are particularly rich in the chemical bonds that release energy.

One might expect a hunting and gathering lifestyle to require more energy than the average western way of life in modern times, but recent research among the Hadza people of northern Tanzania turned up surprising results. The Hadza are still largely hunter-gatherers and tend to be lean and active people. At the time of the study they were obtaining over 95 per cent of their energy from wild foods, yet on average their daily energy requirement (as measured by CO_2 output) was not significantly different from that of American and European men and women, matched for weight and age. This prompted the researchers to suggest that total daily energy expenditure in humans is more a product of our genetic inheritance than it is a consequence of lifestyle.[2]

Nevertheless, individuals do have different energy and nutritional needs, simply because of personal differences even when groups are matched for age, sex, body mass and level of activity.[3] As a rule of thumb, most adults need 2,000–3,000 kilocalories a day so you won't be far off the mark, in energy terms, if you think of people as 100 W light bulbs (100 watts is 2,064 kilocalories a day).

In general, the move towards a settled existence based on agriculture meant that families worked harder and longer in order to grow and process food, while the diet became increasingly dominated by fewer types of plants and animals. Gradually, the transition to agriculture did result in a deteriorating health profile, and not only in the Fertile Crescent.[4,5] A dependence on processed cereals might have been enough to account for increased tooth decay but other health problems resulted from a combination of causes because people were also crowding into bigger settlements together with a growing number of domesticated species. This not only passed disease and parasites from domestic stock to humans, it also served to channel more pathogens from the wild because rats, mice and insect pests also thrived under these conditions.

Health and disease

It is common knowledge that an individual can have plenty to eat but still be critically deficient in dietary essentials—notably one or more minerals or vitamins. The symptoms may be obvious in a living person but are not always detectable or easy to interpret in ancient bones. There are exceptions; vitamin D deficiency, for example, can cause rickets, a disease that is easily diagnosed from bone deformities. The earliest recorded case in the British Isles was a woman who had lived on the Scottish island of Tiree about 5,000 years ago—well within the neolithic. Like others in her community she had not included sea-fish in her diet, but for some unknown reason she must also have lacked sunshine as she grew up.[6]

In skeletons from prehistoric times, the problem of diagnosing illnesses caused by dietary deficiencies is complicated by the fact that malnourished people are also more susceptible to disease so it can be hard to determine the exact cause of death. But technology is constantly improving and museum specimens can sometimes tell us a lot more than they did when they were first collected. For example, using skeletons from mesolithic and medieval times, it has recently proved possible to identify the types of bacteria that lived in the plaque on teeth. It was found that the earlier hunter-gatherers, compared with neolithic and medieval farmers, had fewer types of bacteria that cause tooth decay. The situation is worse in modern (post-industrial) humans as the bacterial mix is less diverse and is dominated by types that cause decay.[7]

In the earliest cities, as in modern ones, wealthy elites would have been eating as they liked but in those days they would have had no more understanding of nutritional needs than anyone else. Nor would they have been immune to diseases—both old and new.

There is no suggestion that human beings, or other primates, have ever been completely free of disease. Even at low densities and eating the diet to which they were best adapted, early humans would have been vulnerable to parasites from the animals that they killed, as well as being infested by lice and exposed to other disease-carrying insects such as mosquitoes. But disease epidemics are not sustained by people who are scattered about in small groups, and some diseases of humans did not even exist in pre-agricultural times because they were acquired later from domestic animals.

The major infectious human diseases mostly have their origins in other species though not all of them have resulted from domestication; some have been contracted directly from the wild. *Falciparum* malaria, for example, is now carried directly between humans by infected *Anopheles* mosquitoes, although the parasite (*Plasmodium falciparum*) probably originated in birds.[8] Mosquitoes cannot remain heavily infected with virulent strains of the pathogen when humans are few and far between so malaria was probably sporadic and without severe symptoms among palaeolithic hunter-gatherers in the tropics.[9]

A single human lifetime can be hundreds of thousands of generations for microbes and during that time they can become transformed by evolution. Each new generation presents a different genome to be tested by natural selection. Cultures of *E. coli* bacteria have been continuously maintained at Michigan State University since 1988—more than 60,000 generations. Since then they have doubled in size, show a higher rate of mutation and have evolved differences in nutrition.[10]

Diseases that pass from other species into humans are known as zoonoses, and the pathogens involved may evolve new adaptations as they make the transition. Five stages have been recognized in the process but not all pathogens progress to stage five. Rabies and anthrax, for instance, are examples in stage two—in which the pathogen passes from another animal to humans in the 'primary infection' but has not been transmitted directly between humans as a 'secondary infection'. Ebola virus is an example of stage three, in which there are secondary infections between humans but they die out after a few cycles. Stage five pathogens have become exclusive to humans and include common illnesses such as mumps, measles, rubella, smallpox and syphilis, as well as *falciparum* malaria.[11]

From a list of fifteen important diseases of temperate zones we find that ten can cause lasting epidemics among large, crowded human populations. Victims may gain immunity but the diseases can persist for years by spreading to other regions, then returning to previous locations when new victims have appeared without immunity. There is reason to believe that at least eight of the fifteen diseases (diphtheria, influenza A, measles, mumps, whooping cough, smallpox, tuberculosis and rotavirus) reached us via domestic animals, and another two (plague, typhus) were acquired from rodents.[12]

Researchers are generally agreed that agriculture, the growth of settlements and the transition to urbanization brought no immediate rewards in terms of health or longevity; instead, they collectively paved the way for longer hours of toil, more malnutrition, and more sickness within social hierarchies that were controlled by increasingly despotic rulers. It does not appear that people were coerced into becoming urbanites, at least not to begin with, so why did people follow this path? As is so often the case, a number of different circumstances and social pressures were probably acting at the same time, and in trying to understand them we must remember that people of the neolithic did not, and could not, have thought about their futures in the way we do today.

A neolithic world-view

Recent research, based on computer modelling, predicts that oppressive social organizations will arise if two conditions are met. First, leaders must emerge who are able to bring benefits, not only to themselves but to their followers, in the form of surpluses that support increasing population. Secondly, followers become locked in to the status quo when the costs of leaving it are seen to outweigh the benefits.[13] Both these criteria would be met, for example, if leadership created unrivalled irrigation systems that provided security and produced crop surpluses for all (as well as accumulating wealth for the leaders) while any opportunities for a better life in the surrounding lands had been diminished or destroyed as outlined at the end of chapter 5.

But there is another important dimension to this that was not covered by the computer model; I'm referring to the persuasive power that leaders could exert in the minds of neolithic people. Most of what we expect our children to learn in primary school was totally unknown in neolithic times. Indeed it can come as a surprise to recall how recently so much has become common knowledge. In the field of health, for instance, when Florence Nightingale, social reformer and founder of modern nursing, died in 1910 she was still unconvinced by the 'germ theory'. She thought dirt was a direct cause of illness, so for her, the talk about airborne germs was a threat to the campaign for better hygiene.[14]

No doubt humans have always tried to explain the world around them, but fundamentally, from the earliest thoughts until just a few hundred years ago, when people tried to interpret what they experienced they could not think in terms of 'the supernatural' as we understand it. Only after natural laws had been discovered and could be expressed in scientific terms could there be any concept of forces or beings that were undoubtedly beyond nature. Until then, nature had none of the boundaries that we now recognize. The world was there to be explained and some explanations would have seemed more obvious and convincing than others, but today's concepts of the natural and the supernatural could not have been sharply distinguished.

At a very early age, people come to perceive meanings and intentions behind the physical actions of others. It is part of becoming a social animal. Research is ongoing, but a number of experiments have demonstrated that infants as young as seven months recognize intention behind the movements of human hands when, for example, they reach for a toy.[15] What's more, infants as young as 10 months can already have come to expect that it will be a human hand that causes something to move, rather than an inanimate object. And this expectation can lead them to infer that a hidden hand is making things move—even when none can be seen. They don't make this assumption if they have already been shown that the thing can move by itself.[16]

As adults, if we see a bean-bag come flying over a wall, we will assume that somebody threw it. But it takes careful experimental design, and a lot of patience, to show that infants can be making similar assumptions even before they are old enough to talk. Now think of those earliest settlers, trying to make sense of their world. We can never know what they really thought but we can be certain that much would have been a mystery at that time. The sun comes up every morning and goes down to somewhere else at night. It might rain or it might be unexpectedly dry for weeks on end. Storms sometimes wipe out crops and months of work, or the weather remains favourable and there is a bumper harvest. Above all, there is the thing that happens to people and other creatures when they cease to show any sort of activity and begin to decay.

Death, like birth, happens to us only once so it has always been a special event, but it has not always been regarded as an all-or-nothing phenomenon. The following passage was written in the late 19th century by Sir Edward Tylor (1832–1917), who was Oxford's first professor of anthropology, and is generally regarded as the founder of cultural anthropology.

> A person who a few minutes ago was walking and talking, with all his senses active, goes off motionless and unconscious in a deep sleep, to wake after a while with renewed vigour. In other conditions the life ceases more entirely, when one is stunned or falls into a swoon or trance, where the beating of the heart and breathing seem to stop, and the body, lying deadly pale and insensible, cannot be awakened; this may last for minutes or hours, or even days, and yet after all the patient revives. Barbarians are apt to say that such a one died for a while, but his soul came back again. They have great difficulty in distinguishing real death from such trances. They will talk to a corpse, try to rouse it and even feed it, and only when it becomes noisome and must be got rid of from among the living, they are at last certain that the life has gone never to return. What, then, is this soul or life which thus goes and comes in sleep, trance, and death ? To the rude philosopher, the question seems to be answered by the very evidence of his senses. When the sleeper awakens from a dream, he believes

> he has really somehow been away, or that other people have come to him. As it is well known by experience that men's bodies do not go on these excursions, the natural explanation is that every man's living self or soul is his phantom or image, which can go out of his body and see and be seen itself in dreams.[17]

Now, I see a recently-published book that combines evidence from religious studies, psychology, anthropology, evolutionary biology and neuroscience to make the case for 'big dreams' being 'a primal wellspring of religious experience'.[18] It is a book that I have yet to read, and I simply repeat the point that I made above: educated people today know the difference between the natural and the supernatural, or at least they can easily find out, but for people of the neolithic there was no such distinction; nature had no known boundaries.

In summary, then, we can say that as infants we tend to perceive meaning and intention behind ordinary events, and from experience we soon learn to make inferences about what is likely to be causing things to happen. Perhaps, like me, you also find the above passage from Tylor to be an entirely plausible description of the way people would have thought about death in the millennia between early settlement and the rise of science. If so, then you will not be surprised by the thousands of vaguely human forms that have been named as being responsible for just about everything that anybody has ever experienced. Which brings us back to those early settlers and whatever it was that drew and pushed them into an ever more urban way of life. I have no trouble accepting the two conditions required by the computer model with which I began this section, but it seems to me that the leaders who rose to prominence in the neolithic would also be those who could most convincingly demonstrate their influence among the gods.

Part two: Onward with technology and economy

8 Economics, human nature, and early limits to growth

Economics, or more precisely, what today we would call political economy, was established as an independent field of study in 1776. In that year the Scottish philosopher, Adam Smith (1723–1790) published *An Inquiry into the Nature and Causes of the Wealth of Nations*. It is a hefty book (originally six separate volumes) and economic theorists have found it easy to be selective and to quote bits of it in support of one view or another, which they claim to be Smith's way of thinking.

In particular, Smith is often cited as favouring a free market economy because he reasoned that when all people are free to pursue commercial self-interest there are benefits to society as a whole. Market prices will reflect the laws of supply and demand because prices will not have been distorted by manipulators. These days it seems almost impossible to find any reference to Adam Smith that does not mention his metaphor of an 'invisible hand' that serves to direct self-interest towards the common good, at least in some circumstances. One might think that it was Smith's favourite phrase but he actually used it only once in the *Wealth of Nations*—to remarkable effect in terms of the interest it generated.[1]

Smith begins the *Wealth of Nations* (as it is now usually called) by explaining how economies have been transformed by machinery and the division of labour. He had been impressed by a small business which could turn out more than forty-eight thousand pins in a day from the work of ten men—the equivalent of one person crafting 4,800 pins over the same time period. He then reflects upon the origins of such progress and concludes that it is not the result of human wisdom so much as our 'propensity to truck, barter, and exchange one thing for another'. In other words, Smith was convinced that the division of labour in humans had grown inevitably out of our inclination to trade.

Whether people are born with a trading disposition, or whether it comes from an underlying covetousness, Smith was not so sure but he was confident in saying that it had not arisen in any other kind of animal.[2] In this he was quite right. Pairs of chimpanzees can be taught to use tokens for trading with each other but the trading idea has to be taught and kept going by humans.[3]

Smith's emphasis on self interest as motivation might indicate a very hard-headed view of human nature but there is much more in his writing that shows a concern for fairness and for the welfare of society at all levels. He was especially critical of what has been called 'crony capitalism' at a time when trade unions were not even legal. Masters, too, said Smith, 'sometimes enter into particular combinations to sink the wages of labour'.[4] Nevertheless, fair or otherwise, in Adam Smith's view all market economies evolved out of human self interest.

We will meet one or two other eighteenth century economists in the next chapter but here I need to visit the twentieth century for an entirely different view of human nature—that of Karl Polanyi (1886–1964). Polanyi had a passionate interest in the social, cultural and political history of economics but he is not easily slotted into any category of economists. Son of a Hungarian father and Russian mother, Polanyi studied law in Hungary but lived and worked in Austria, England and America before finally settling in Canada. He was highly critical of free market doctrines but was not a Marxist and his ideas, in his time, were very much his own. In more recent years there has been a revival of interest in Polanyi, especially among sociologists.[5] The Karl

Polanyi Institute of Political Economy was set up in 1988, thanks partly to the efforts of Polanyi's daughter, in affiliation with Concordia University in Montreal, Canada.

Polanyi's best-known publication is the book *The Great Transformation: The Political and Economic Origins of Our Time*, first published in 1944, and last published, with corrections, in 2001.[6] The book grew out of Polanyi's researches for courses that he was teaching on English social and economic history. By now you may be wondering what all this has to do with the neolithic and the economies of antiquity. I hope to make that clear in a moment.

While acknowledging Adam Smith as a great thinker, Polanyi dismisses his suggestion that the division of labour arose from man's disposition to barter. According to Polanyi 'that propensity had hardly shown up on a considerable scale in the life of any observed community, and had remained, at best, a subordinate feature of economic life.'[7] This was a big issue with Polanyi because, in his opinion, the nineteenth century economists who wrote in favour of the free market system had all been misguided in thinking that bartering and trading for personal gain was perfectly natural behaviour. Not so, Polanyi insisted, which is why in city states and feudal systems alike 'the production and distribution of goods is organized in the main through collection, storage, and redistribution, the pattern being focused on the chief, the temple, the despot, or the lord.'[8]

In Polanyi's opinion, humans are neither capitalists nor communists by nature; their economic concerns are essentially social. Man's economy, asserts Polanyi, is as a rule submerged in his social relationships. He does not accumulate material goods out of self interest but to 'safeguard his social standing, his social claims, his social assets. He values material goods only in so far as they serve this end.[9]

It is a question worth bearing in mind with regard to our modern consumer society. Given the range of cultural and individual differences that exist in today's world, are people motivated more by social standing or simple covetousness? Are we aware of it? And does it matter? It obviously mattered to Polanyi and his followers because it was central to his interpretation of economies both past and present. Polanyi was insistent that all economies before the nineteenth century were an integral part of prevailing cultures. 'Embedded' was the term he used, expressing the idea that ancient economies were not autonomous but were instead 'subordinated to politics, religion, and social relations'.[10]

This last point is not easily refuted or confirmed because markets can have some degree of autonomy, and economic sociologists it appears, can argue indefinitely about the extent of different forms of 'embeddedness' in different cultures and at different periods of history.[11,12] But it brings us back to the question that was raised in chapter 6 with regard to Uruk: to what extent were economies centrally controlled in those first cities, dynastic city-states, and later civilizations? Almost entirely, if we believe Polanyi.

But archaeologists and those who decipher cuneiform scripts have learned a lot more since Polanyi's time as his critics are pleased to point out. The economist Morris Silver devoted a substantial part of an entire book to refuting Polanyi's arguments. Private enterprise, it appears, has been alive and well since the first deals were recorded on tablets.[13]

So how are those early economies best described? Without doubt, commercial and many other activities were first centred in the temples. A city's main temple was dedicated to the city god while other gods and goddesses were located and worshipped in numerous other temples. The buildings had management staff and provided physical space, including granaries, for storage so that contributions from the community could be collected, stored and redistributed. Because harvests are not consistent from year to year, some citizens would be unable to make their contributions on schedule so arrangements for credit also originated in the temples, as did loans with interest.[14,15] Debt, it appears, is as old as capitalism.

There is ample evidence that sales, loans and deals of all kinds were also conducted by individuals acting in a private capacity, not only on behalf of the temple. Morris Silver, referring to ancient Greece and early antiquity in general, sums it up this way: 'The direct evidence for trade, occupational specialization, supply-demand-determined prices, investment in human and material capital, coinage, banking, and other modern phenomena is uneven with respect to time and place but is, nevertheless, abundant.'[16]

The word 'coinage' does not refer to coins as we know them. From Mesopotamia we have the earliest records of payments being made with weighed and sealed bags of silver, like purses of known value that served as high denomination coins. According to the British Museum, the earliest coins were made about 2,700 years ago in the ancient kingdom of Lydia in modern Turkey. The metal is a natural alloy of gold and silver called electrum.

Returning to more ancient times, there were patron deities for artisans and merchants, and the gods played an economic role as witnesses and guarantors of honest dealing. .Transaction costs (the costs of doing business) were high because the means of communicating were so limited and transport was primitive, making trust all the more important. The confidence inspired by temples must have been invaluable. In Greece, as well as in Mesopotamia, the sun-gods, in particular, were approached to witness the swearing of oaths because they saw everything that occurred, both in the land of the living and in the nether regions where they went nightly.[17]

From temples to banks

As royal dynasties increased their military power through the Bronze Age, the roles and identities of kings and gods became blurred and palaces gradually took over the functions of temples. In Europe the rise of an Iron Age village culminated in the Roman Empire—an empire that lasted under Rome for 500 years and developed an immense economy, largely based on slave labour and with military spending taking most of the imperial budget. Through those centuries the central, redistributive, economy waxed and waned together with a good measure of private commerce, driven by profit, and overseen by a pantheon of a dozen major gods and hundreds of subsidiary ones.[18,19] Under Rome the banking practices developed by the Greeks also grew in scope and sophistication, though banking was still done by independent money-changers. Debts could be discharged by making deposits to the money-changers and the transaction would be registered by public notaries.

In the Roman Empire, as in all agrarian economies in pre-industrial times, the value of production per head of population was very low by modern standards but because of the geographical size of its territory Rome was less vulnerable to the vagaries of weather and the impacts of short-term climate change than were smaller economies. In any case, the consensus seems to be that the Empire came to maturity during a favourably warm climatic period and its eventual decline had a combination of interacting causes—the ravages of war and epidemic diseases being especially important among them. The year 476 is usually taken to mark the end of it because it was the year in which a former German warrior, Flavius Odoacar, took control of the western half of the Empire and became the first barbarian king of Italy.[20]

The power of the Christian Church increased and trade diminished after the Roman Empire collapsed. The Church took a dim view of charging interest on loans, which they called 'usury' and that attitude, at least for a time, worked in favour of Jewish moneylenders and rich borrowers. In Italy, however, during the 13th and 14th centuries, moneylending was increasingly taken over by wealthy Christian families, especially from Florence, whose creative accountants recorded interest by another name. A series of family dynasties became phenomenally rich and influential by financing the feudal powers of Europe—including the papacy.[21]

Paper money began to circulate in the 16th century in the form of receipts that goldsmiths had issued in return for gold deposits. Some receipts would promise to pay not just the depositor but the bearer of the receipt, which made the notes transferable to some extent. Individual moneylenders were gradually replaced by private banks, and by the early 17th century the banks were providing a commercial service for more ordinary citizens as well as the elite. The first central bank (bank for bankers) was Sweden's Riksbank, which opened under parliamentary control in 1668 and is the oldest surviving central bank. The Bank of England was set up as a private bank 1694 to raise funds for a cash-strapped government at war with France, but it went on to become a central bank long before it was nationalized in 1946.[22]

Ecology and early limits to growth

So far in this chapter I have skipped lightly through a few thousand years with some thoughts about the origins of economies and economists' views of human nature, but in reality there were many non-human limits to economic growth and the scale of urbanization, and some of these limitations are far less speculative than are opinions about human nature.

Estimating population sizes and densities from times long past is notoriously difficult, but indicators from different sciences have been put together and scrutinized by specialists over a long period so there are some generally-accepted figures. From six or seven million people in the early Holocene the world population had reached about a quarter of a billion by the start of our western (Gregorian) calendar—the year one. The population had increased to three quarters of a billion by the middle of the eighteenth century, which can be thought of as the eve of the industrial revolution.[23]

The rate of population growth shows variations in time and place resulting from the interactions of wars, changes in climate, and disease, but overall it was disease that prevented more rapid population growth in agricultural communities throughout the world. Major epidemics are well known from historical times; the Justinian plague, first recorded in the sixth century, is believed to have killed more than 25 million people over a period of 200 years and may have reduced the world population by as much as 20 per cent. The Black Death, after reaching Europe in the late 1340s, had a major impact on that continent for several decades, although estimates of total mortality range from less than one third to 60 per cent of the population.[24,25] Both plagues were caused by different strains of the bacterium *Yersinia pestis.*

The distribution of people around the world had not changed substantially by the year one. Most of the world, including the Americas, Central Asia, Northern Europe, Siberia and Australia, were still without people or were very lightly populated with no large settlements. The human impact on natural ecosystems was therefore geographically limited.

Most wild ecosystems are completely destroyed where they are replaced by fields of crops, and they can be drastically changed when the land is heavily browsed and grazed by herds of livestock. In terms of geographical extent, the conversion of land to agriculture is the biggest single impact that humans have had on the wild, and from the very beginning, the impact appears to have spread beyond the areas immediately affected by the farmers.

In an exhaustive statistical analysis of prehistoric animals and plants it was found that over a period of 300 million years an average of 64 per cent of species always occurred together with another species—as in predator and prey or two types having very similar niches. The analysis concentrated on North America, where agriculture began about 6,000 years ago. Concurrent with this change in land use the percentage of associations between species dropped to an average of 37 per cent. As the senior author of the report explained: 'Humans have broken up plant and animal populations by destroying and fragmenting habitats. Their ranges are smaller, and no longer overlap in the way they once did.'[26]

Habitat destruction and fragmentation have always been a consequence of agriculture but until a few centuries ago there were still vast regions in the world that were relatively unaffected. In settled parts, on the other hand, some effects of human activities were as ruinous in early times as they are today, and not only to wild nature but to agriculture as well. The huge irrigation systems in Mesopotamia acted as evaporation pans and gradually increased the salt content of the water until wheat could not be grown. Barley is more tolerant of salt but eventually farmers struggled even to grow barley.[27]

Deforestation in Mesopotamia, ancient Greece and especially in the Mediterranean basin during the decline of the Roman Empire, arguably added soil erosion to other ecological and economic problems.[28] I say 'arguably' because terracing of slopes was widely practiced to prevent erosion of farmland and we know that terraces break down when they are neglected or abandoned rather than when they are worked more intensively, but we are still talking about human activity to begin with.

The need for energy not only grew but became more concentrated with settled agriculture and the growth of cities. Whereas hunter-gatherers had needed only a few thousand kilocalories a day from food and firewood, now there was livestock to be fed, and timber to be found for constructing buildings, fences, carts, boats and so on. Fuel was required not just for cooking and keeping bodies warm, but for heating buildings and firing kilns—first for pottery then for metal smelting and forging. All the fuel came from wood and other plant material that is collectively referred to as plant biomass or phytomass.

Most forests store no more than about 200 tonnes of phytomass per hectare (mostly in tree trunks) and take decades to regrow if they are clear-felled. On that basis it has been calculated that a pre-industrial city, depending on severity of climate, would need access to between 50 and 150 times its area as a source of phytomass. The need could be as much as 250 times the area of the city if a proportion of the wood was first made into charcoal for kilns, so the energy constraint alone would have limited the size of cities in pre-industrial times.[29] New sources of energy, the next big turning-point in our story, were key to further expansion.

9 The coming of industry:

Among my recollections of childhood in Lancashire is the memory of being woken on cold, dark mornings by the clatter of mill girls' clogs as they hurried over pavements and cobblestones to the cotton mills.

As a young boy I was allowed into a mill on one occasion to deliver something to my aunt. I don't remember what it was but I have not forgotten the experience. Once inside the grim brick walls, I was led across what seemed like an acreage of floor space between rows and rows of clanking, hammering looms from which the young women were wresting a living. My aunt, I was told, was among those who earned extra pay because of the number of those deafening machines that she could work simultaneously. And like all the experienced weavers, she could gossip, with those who were close enough, by means of an in-house combination of sign language and lip-reading.

The textile industry was entirely transformed by mechanization, and it became the flagship of the British economy, a major triumph of the Industrial Revolution. In the mid nineteenth century, textiles made up more than 40 per cent of British exports and at that time the notion of importing cotton cloth would have been unthinkable to workers in the north of England; they were meeting half the world's demand for cotton goods. But at the time of my visit to the weaving mill, in the late 1940s, the industry was already in decline, and a couple of decades later, as the end approached, mill closures were being announced almost weekly.[1,2]

Substantial libraries could be filled with books and articles on the Industrial Revolution. The main events are all well documented although historians have different opinions about why they first occurred in Great Britain. I can add nothing new to the well-worked topics but the coming of industry had an impact even greater than that of early agriculture on the world's ecology and economy so it cannot be left out of this narrative. And because people have been making cloth for longer than they have been making pottery, there is no better industry upon which to focus.

Unlike pottery, textiles are perishable so archaeologists find very few examples from early times. Fabrics woven from flax fibres (linen) are known from Çatalhöyük, and a small fragment, wrapped around a tool made from antler, has been found at another site in Turkey and radiocarbon dated at nearly 9,000 BP.[3]

The idea of weaving long strands to make cloth presumably developed from netting and basketry so could have pre-dated agriculture by a very long time. Twisted flax fibres, found in clay from a cave in the Republic of Georgia, are believed to be more than 30,000 years old.[4] They are extracted by partially rotting ('retting') flax stems in water, which not only softens them but turns them a golden colour—hence the term 'flaxen-haired'. The language of making cloth has long been part of our wider vocabulary—you may have been on tenterhooks at some time, or spinning a yarn as you were weaving through traffic.

The wild ancestor of domesticated flax occurs naturally from southern England to western Asia so it was available throughout the Fertile Crescent. The domesticated form is now grown worldwide for its stem fibres but more so for the linseed oil obtained from its seeds (the flax family is Linaceae). There are other plants with fibres similar to those of flax. Hemp, jute, and nettle fibres have all been used for thousands of years.

At its simplest, there are two main processes in making cloth:

spinning and weaving. First the fibres need to be twisted or spun together to make threads, and then the threads must be woven into a fabric. In the weaving process a number of strong threads (the warp) are tensioned in a row of straight lines on a frame, then another long thread, (the weft) is woven around and between them. It is easy to imagine that the people who first attempted this started out with a frame of four sticks but they are likely to have got beyond that stage quite quickly. The earliest loom was probably the horizontal ground loom, in which a frame is held above the ground on four pegs. But the loom used throughout antiquity was set vertically with the warp hanging from the top beam and tensioned by clay weights tied to the warp at the bottom. Weights are often all that remains for archaeologists to find so they cannot always be sure that the weights were used for looms. The technique is kept alive in many countries by individual weavers and hobbyists.

Wool was probably being woven in the same way as other fibres soon after it became available—which largely depends upon times of domestication and selective breeding for wool. Cuneiform tablets from Uruk refer to some flocks of sheep as 'wool sheep' but yields were low by modern standards.[5] Modern breeds can produce so much wool that they are unable to survive as wild animals. A lost Merino sheep did survive near Canberra, Australia, for nearly five years, and when it was found it was tottering about in a fleece that, at 40.45 kg, was close to half the animal's total weight![6]

Silk, obtained by unwinding a silkworm's cocoon, provides threads hundreds of metres long and strong enough to be handled. The Chinese were probably perfecting secret techniques with silk in the early Bronze Age but smuggling silkworms out of China was a crime that carried the death penalty so sericulture was not established in the west until well within Roman times. Before that there was limited use of silk outside China from wild species of silkworm. Cotton has an entirely different history because different species of cotton plants (*Gossypium*) are native to tropical and subtropical regions across the world. All these different fibres, were made into cloth, and it is probable that women spent more hours spinning and working at their handlooms than were needed for food production and pottery combined.[7]

Energy and machines

By medieval times the textile industry had become central to the economies of towns, with increasing demands for everything from embroidery to the sails of ships. There was no such thing as cheap clothing. Of course the towns of Europe had different specializations and by that time there had been many advances in tools and technology. The types of fibre (wool, silk, linen) also required quite different preparations and treatments but, as far as one can generalize, the making of cloth remained a cottage industry until the eighteenth century.

Basically, merchants had become employers by distributing raw materials to households, where families would own or hire the necessary tools and organize their own time and labour to bring the raw materials to an agreed stage. The merchant would then collect it and take it elsewhere for the next stage, or else sell it if it was finished. The arrangement came to be known as the 'putting out' system and was not only used for textiles. It still exists in Asia and South America. Millions of people in India, most of them women, are employed in this way and are known as dependent sub-contract workers or homeworkers.[8] Most of us are probably familiar with modern western versions of outsourcing.

Thinking about the system historically, it is easy to imagine needy workers being ruthlessly exploited by contractors, and the contractors having difficulty meeting deadlines and maintaining quality. The downside of the division of labour is the cost that can result from errors made late in a sequence of operations. This may not have been obvious in Adam Smith's time, but when rows of machine operators began mass-production it must have become nerve-racking to those who lost most from mistakes or carelessness. As a social concern, the introduction of powered machines and the move from staying at home to working in 'manufactories' is inevitably seen quite differently by Marxists and capitalists. Writers in neither group have been particularly concerned with ecology

Powered machinery obviously had to be located where a usable form of energy was available and in the eighteenth century there was not a lot of choice. People first began to control energy beyond their own bodies when they got the hang of throwing things and starting fires. Later, they learned to use the wind on a sail, the potential energy in a pulled bow, and the muscles of domestic animals. By the time of the ancient Greeks, windmills and waterwheels were also widely used and that just about completes the options available to Richard Arkwright in 1769, when he patented the world's first powered spinning machine that could make strong warps for cotton.

In England, Arkwright and his partners first built a small factory in Nottingham where the machines were powered by horses walking in a circle and turning a capstan, but by all accounts it never became fully productive. In 1771 Arkwright moved to Derby and started a factory that used water power from a tributary of the River Derwent; there had been a corn mill there since the thirteenth century. Arkwright's factory did well and he was soon building factories in other parts of England and as far afield as Scotland—all located near waterways that could be used for turning waterwheels. So began the factory system, although there were many more small concerns than big textile mills for a long time to come.

As the work could not be distributed to the workers, the owners of big factories brought their employees to the workplace by building local housing. Large families were preferred because they provided more workers per house. In the early decades of production the mills were in operation day and night and a large proportion of the workers were young children.[9]

While inventive minds were coming up with more efficient machines for making textiles, there were those who were trying to develop new ways of providing power to other kinds of machinery. For all fuel other than food, people remained mainly dependent on phytomass until the seventeenth century, despite coal being readily available on the surface in some places, and having been extracted from shallow mines for millennia.

Coal, like oil and natural gas, represents the energy fixed by photosynthesis in times long before humans existed and the first people to use it could have had no idea what it was. At least one group of villagers in north-west China were burning black stones as fuel more than three thousand years ago, though the discovery that coal would burn could have been made much earlier; it would have been a simple matter for somebody to include a lump of coal among the hearth stones. But not by chance would anybody ever discover a way of converting the heat from burning coal to mechanical energy, and yet without the energy from fossil fuels the Industrial Revolution could not have continued.

The power of steam was harnessed in 1698 with a vacuum device for pumping water from mines but it was another 85 years before there was a piston engine using steam pressure to drive rotary machinery in factories. And it took decades more (partly because of patent restrictions) to develop a steam engine that could replace horses for transport. At first the engines were used for hauling freight around mines and ironworks but Stephenson's famous *Rocket* demonstrated its worth in 1829 and the world's first inter-city railway, from Manchester to Liverpool, opened in 1830. The original locomotive is on display at the Science Museum in London.[10]

Industry and revolutions

The Industrial Revolution, as a period in British history, is generally understood to cover the last quarter of the eighteenth century during which the textile industry, and much of British society, was transformed by mechanization and urbanization as outlined above. It has no agreed beginning or end but the term is so familiar and convenient that writers commonly try to explain what they mean rather than avoid the term. An editorial leader in *The Economist*, for instance, refers to it as the first industrial revolution with the second one beginning early in the twentieth century 'when Henry Ford mastered the moving assembly line and ushered in the age of mass production'. The author of this article argues that a third industrial revolution is now underway as manufacturing goes digital and completely new processes, such as 3-D printing, replace the old assembly lines.[11]

The founder of the World Economic Forum, Klaus Schwab, counts the revolutions rather differently. In his view the third revolution began at the end of the 1960s with electronics and IT but emerging technologies such as 3-D printing, robotics and nanotechnology constitute the beginning of a fourth revolution.[12] The European Textile Network, based in Germany, has yet another interpretation in which the eighteenth century developments in England are regarded as a second industrial revolution; the first being the development of cottage industries in Europe that began prior to the thirteenth century.[13] Every continent could probably define its own industrial revolution differently so to avoid confusion I will use initial capital letters only to refer to that particular period of British history.

Britain tried to prevent the export of new technology and trade secrets for as long as possible, but it soon became a lost cause so policy was reversed and machines and technology became the new exports. Belgium, having an established textile industry and good coal reserves, was quick to industrialize but France remained an agricultural economy for rather longer. Germany, rich in coal, iron ore and scientific expertise, outstripped Britain in steel production and was the world leader in the chemical industry by the end of the nineteenth century.[14] Meanwhile, America was forging ahead with great efficiency, adopting a high-volume, low-margin approach to products and services as millions of immigrants swelled the home market. Before the turn of the century the USA had overtaken Britain in manufacturing output.[15] At that time Japan's industrial revolution was just taking off but by 1973 its output of goods and services was second only to America's.[16] I haven't even mentioned the rise of China or the four Asian Tigers—Hong Kong, Singapore, South Korea and Taiwan—but I have moved well ahead of the Industrial Revolution so it's time we went back to take stock.

Economy and ecology in 1800

There is nothing critical about the year 1800 but it will serve to mark the period when much of the world was on the threshold of economic change but before industry had made its mark on a grand scale anywhere. Until then the productivity of nations was mainly dependent on the number of producers. In other words, the gross domestic product (GDP) of nations was more or less a consequence of its population size so that two nations, India and China, together accounted for roughly half the world's GDP. Moreover, in terms of buying power, average earnings were much the same as they had been for thousands of years. About three quarters of the income of European farmworkers was spent on the food that would most cheaply meet the energy needs of the family and yet they were earning more than those of India, China and Japan.[17]

Agriculture had raised the human carrying capacity of land to a much higher level than existed when the land was supporting natural ecosystems, but land and crops must still have been setting the higher limits, at least in theory. How many people could an acre support, given all the different climates and crops. It was the sort of question that troubled an Anglican cleric and scholar called Thomas Robert Malthus (1766–1834). Malthus had been privately educated before entering Jesus College, Cambridge, and had studied widely but ecology did not then exist as a formal subject. The language of economics was still that of Adam Smith, and energy was only vaguely understood in the sense that it was some sort of 'living force'. Nevertheless, Malthus had done well in mathematics and he recognized that populations increased exponentially (or geometrically, as he called it) but would eventually be limited by the productivity of the land, which could only be increased arithmetically. Upon reaching the limits of productivity, population growth would cease because, as he explained in his *Essay on the Principle of Population* in 1798, 'the poor consequently must live much worse, and many of them be reduced to severe distress ... during this season of distress, the discouragements to marriage, and the difficulty of rearing a family are so great that population is at a stand.'[18]

Malthus had reason to think that human populations could potentially double every 25 years (in fact they can double faster than that) and although he knew that the productivity of farmland could be increased over time he found it inconceivable that it could be increased exponentially to keep pace with the repeated doubling of populations. 'Where is the fresh land to turn up?' he asked. 'Where is the dressing necessary to improve that which is already in cultivation?'[19] His reasoning was sound but like so many who have tried to predict the future, he failed to imagine new developments, such as the ways in which fossil fuels would provide energy, and advances in contraception. In any case, we know that Malthus would have disapproved of modern contraception. He remarked that 'it is difficult to conceive any check to population which does not come under the description of some species of misery or vice.'[20]

Malthus was prompted to write his *Essay* mainly as a response to some wildly optimistic views of future society that were being expressed in the spirit of the French Revolution (1789–1799). It was in no sense a weighty publication (it is frequently referred to as a pamphlet) and Malthus had it published anonymously, but he was soon identified and then found himself the centre of quite a fuss. The prevailing view was that an expanding population could only be a good thing and now he was challenging that view with very little contrary evidence. After all, people were not constantly starving to death, except during famines when the cause was usually known.

Worse still, it was being claimed that the 'poor laws' in England were increasing the number of malnourished babies among the poor rather than relieving poverty. Malthus accepted that he had little numerical evidence to support his claims but he travelled to Europe to gather more facts, and he continued his writing which culminated in a sixth and much heftier version of his *Essay* coming out in 1826. At this point I must mention another British economist, David Ricardo (1772–1823).

On the face of it, Malthus and Ricardo had very little in common. Malthus' parents were minor English gentry while Ricardo was born to Jewish immigrants. Moreover, the two economists first met face to face in 1811 as a result of disagreements over economic policy, and they continued to meet and disagree about economics until Ricardo's death. But Ricardo was supportive of Malthus' claim that unchecked population growth must outstrip food supply and the two remained very close friends. More to the point, their publicized arguments were building on the work of Adam Smith but taking a more systematic approach and laying the foundations of nineteenth century classical economics. It has been remarked that the friendship of Malthus and Ricardo 'was, very likely, the most remarkable and most fruitful collaboration in the history of economics'.[21]

Malthus would surely have been amazed if he could have known that economists would still be troubling themselves over his first pamphlet and 'Malthusianism' more than 200 years later, though much of what is claimed in his name today goes far beyond what he ever said.

Leaving economics out of it for a moment, anybody who knows the basics of ecology (or has only read chapter 2 of this book) will understand that Malthus had simply recognized the principle of carrying capacity. Populations of early humans were limited by the productivity of their habitats, which is why population densities varied according to territory. But with division of labour and the distribution of food, the basics of ecology were overlain with all the complications and politics of particular distribution systems. Karl Marx came vaguely within sight of it when he declared: 'An abstract law of population exists only for plants and animals, and even then only in the absence of any historical intervention by man.'[22] He was not likely to have given it much more thought because biological limitations don't fit well with communist ideology; indeed, as late as 1948 the 'reactionary idealist theories' of western genetics were officially rejected in the Soviet Union.[23]

So why has Malthus remained controversial for centuries? I believe there are two main reasons. The first is that from urbanization until well past the Industrial Revolution, disease held back population growth and generally prevented it from reaching limits that might be set by farm productivity. This allowed economists to continue arguing about whether or not food production could ever be limiting. Secondly, by the time the main killer diseases had been brought under control, technology was raising new worries about the pressure of population on other resources and the new concerns were swept together with the original Malthusian argument.

Disease, as described in the previous chapter, continued to limit population growth as it had done since urbanization, but overall the population graph continued to meander upwards as permitted by agriculture. Global population in 1800 probably stood slightly below one billion but life expectancy at birth (see chapter 2) was probably still less than 40.[24] High fertility and high mortality, especially infant mortality, was still the prevailing pattern in the eighteenth century. Richard Arkwright was the youngest of thirteen children, only seven of whom survived childhood. Karl Marx was one of nine children, Malthus one of seven, and Ricardo one of seventeen.

Surprisingly little progress had been made in disease control by 1800. We were taught in school that the ancient Romans and Greeks were a hygienic lot who had baths and proper toilets at a time when the British were wallowing in filth, but recent research indicates that the Roman fondness for running water would have done little to limit the spread of disease. Not even the poshest privy in antiquity would have had an S-bend, so flies, cockroaches and anything else small enough would have been able to reunite people with their waste. And proper sanitation is not likely to bring about widespread improvements in health unless it is being used by about three quarters of a population.[25]

Nonetheless, the population in 1800 had been reached after an estimated doubling time of only about 220 years, which was rapid growth compared with that of earlier millennia. One billion (a thousand millions) is an impossibly big number to imagine but it may help to think of it in terms of something familiar—like the ticking of a clock. If you had a clock that showed the seconds and it were possible to keep on counting them you would be 31 years and eight months older by the time you reached a billion!

Asia had about two thirds of the world's people in 1800 but the Americas were burgeoning as a consequence of immigration.[26] Globally, the conversion of wildlands to farmland was now occurring at an unprecedented rate in many parts of the world. Natural forests were being cleared both for fuel and building materials in addition to the expansion of farmland. This had been ongoing for as long as humans had been settling, but fuel for making iron in blast furnaces was now in demand as never before. Europe was being deforested.

Charcoal has the same energy density as good quality coal so coal-mining replaced tree-felling only after the remaining trees failed to meet commercial needs. In England this had occurred by the mid-seventeenth century.[27] Several other European countries are said to have reached their lowest point in forest cover by the early nineteenth century.[28] This seems to suggest that forest cover began to return later but what it really means is that tree plantations are being classed as forest. Ecologically, they are not the same thing at all, especially in warmer climates. At high latitudes, natural forest can be made up of uniform stands of conifers, but where winters are less severe the woodlands show much more variety because a wealth of plants and associated animals come together by natural processes over very long periods. Managed plantations of trees, especially as production forests, have more in common with agriculture.

In earlier times deforestation was clearly linked to human population density but the link became less obvious with increasing global trade and new methods of transporting natural resources to distant centres of population. In 1800 there were still vast regions in the world that were barely affected by human demands but the entire world was on the threshold of profound ecological change. For another century or so the world would continue on its upward economic trajectory. The growing population was increasingly supported by industry—which was now entirely dependent on energy that had been fixed on earth by the photosynthesis of an earlier era.

10 Into a different world

Steam power brought change to all levels of society but the nineteenth century was a time when everything, including society itself, continued to change in completely new and unpredictable directions. Cause and effect are not always obvious or direct but there were so many discoveries and inventions in the nineteenth century that it seems fair to say that social change resulted mainly, though not entirely, from new technologies. It was the century that gave us steam engines, railways, gaslight, electricity, refrigeration, the telegraph, the internal combustion engine, phonographs, telephones, microphones, vaccines, use of x-rays, anaesthetics, photography, typewriters, calculating machines—and a lot more.

I said it was the century that gave us these things but they were not all immediately available to ordinary working-class families. Most houses in America had electricity by the end of the 1920s but European homes were far behind the USA in that regard. Still, the nineteenth century, even for nations that were lagging behind in technology, was a period when entirely new directions in life were opened up, different world-views became possible, and ordinary people became aware of such things.

Faith and reason

To put this in context I need to refer to a more distant past. The early Middle Ages in Europe, say from the fall of Rome in 476 through the tenth century, is sometimes referred to as the Dark

Ages, either because it is poorly documented, or in a derogatory sense because there is so little intellectual light to be seen. But modern historians, if they choose a label at all, are inclined to be cautious and call it the Migration Period rather than the Dark Ages. Unarguably it was a time of violent upheavals as the tribes of northern and central Europe, the so-called barbarians, battled their way through what had been Roman territories and kept redrawing the socio-political map.[1]

Through the remainder of the Middle Ages, from the eleventh century onward, a more stable and mainly Christian Europe prevailed. Dates are somewhat arbitrary but some historians conveniently mark the end of the Middle Ages by the fall of Constantinople to the Muslims in 1453. This put an end to the Byzantine Empire, which had lasted for nearly a millennium beyond the fall of its Roman ancestor. Other historians favour a looser definition in which the Middle Ages, or medieval period, extends until that revival of arts and literature known as the Renaissance, which began in Italy in the fourteenth century but was still in progress in other parts of Europe by the sixteenth.

There are a few other popular 'Age' labels that need to be clarified. In the simplest and most sweeping terms, the late Middle Ages is said to represent an Age of Faith in which the true facts of life were known only to God, and mortals could discover them only by interpreting the scriptures. Scientific inquiry in Europe was, in effect, Bible study. In this view, the 17th century became an Age of Reason that led to the Enlightenment of the 18th century and the cascade of scientific and technical discoveries that came with it.

Yet it was as late as 1633 when Galileo Galilei, under threat of torture, famously recanted his opinion that the earth moves around the sun. Nonetheless, he spent the rest of his life under house arrest. Thirty-three years earlier, Giordano Bruno had been burned at the stake for his cosmological insights. Biology at that time was less contentious than astronomy, provided that plants were studied for their use as food, medicine or poison, and animals were portrayed in ways that supported biblical references to their attributes. Some completely mythical creatures, such as the satyr, basilisk and unicorn, were included in medieval bestiaries, and there were some wildly inaccurate accounts of animals that really did exist. For example, in a book by Edward Topsell, first published in 1607 and reprinted fifty years later, weasels are said not to 'engender nor couple in their hinder parts, like other four-footed beasts, but at their ears, and bring forth their young ones at their mouth'. Many animals were also believed to have curative properties that were entirely fictional, as in the following little gem: 'Sodden mice are exceeding good to restrain and hold in the urine of infants or children being too abundant, if they be given in some pleasant or delightsome drink.'[2]

Nonsense of this sort was perfectly acceptable but if medieval intellectuals put forward anything that was contrary to the clerical interpretation of the scriptures they risked being charged with heresy. The consequence of this was summarized bluntly by the late John Moore, a biologist, when he wrote that 'the attitudes of the Church prevented the development of science for more than a thousand years and inhibited it for centuries thereafter.'[3] Arab scholars, for a time, had access to the science and philosophy of ancient Greece and were free to use it as a basis for their own learning but by about the 13th century the Arab world was under the influence of powerful caliphs who, like medieval Popes, opposed any free inquiry that might distract from interpreting God's purpose. Arab learning then became more like Christian scholasticism in scope, and it can be argued that Arab science has still not recovered.[4]

The mythical beasts and fanciful tales of the medieval bestiaries had largely passed into folklore by the eighteenth century, by which time the widespread interest in gardens and herbal cures had produced the first attempts since Aristotle to bring order to chaotic lists of thousands of different plants. The Swedish naturalist, Carl Linnaeus (1707–1778), was by no means the first to attempt a classification but he is immortal in biology because he was consistent in establishing the system of using two names (genus and species) for all plants and animals. We still use his binomial system but his scheme of classification had little resemblance to the present one. For a start, living organisms are now separated into six kingdoms whereas Linnaeus recognized only three kingdoms in nature—animals, plants and rocks.

Linnaeus was deeply religious and regarded all things in nature as evidence of God's design. He would never have wanted to seem deliberately lewd but in classifying plants his reliance on the details of sexual parts really did upset the social norms of his time.[5,6] As long as flowers were delicate things of fragrance and beauty they could be regarded as the essence of femininity and the perfect subject for ladies' art and study. But if they had to be seen as naked genitalia they were altogether something else. And once the idea was accepted there was no shying away from it; studying plants involved examining stamens and styles and talking about sex. One flower in the pea family resembles a human vulva sufficiently for it to be described as such in the vernacular of several languages. Linnaeus duly established the genus as *Clitoria*. The species referred to is *ternatea*, simply named for the place where the type specimen was found. But if the Linnaean method of classifying plants caused some upset it was nothing more than a brief flutter compared to the furore about evolution.

Evolution

For much of the nineteenth century the prevailing world-view was that living things had not changed since they were first created by God a few thousand years earlier. That is not to say that every single person was content to think the same way, with no doubts or questions ever having been raised. It was common knowledge that selective breeding of crops and livestock could produce change, and that hybrids differed from both parents. Change was obviously possible under domestication but continuous change, happening through all time in the wild, was a different matter.

As is often the case in human affairs, new insights emerged from the efforts of a few individuals. A Scottish geologist named James Hutton (1726–1797) published his conclusions that the earth had been shaped and reshaped by natural processes, not catastrophes of the biblical sort, and they had taken an unimaginable length of time. Another Scot, Charles Lyell (1797–1875), built on Hutton's work and was largely successful in convincing the scientific community of the truth of it. So Charles Darwin (1809–1882), had all the credibility he needed with regard to the time component of his theory of evolution, although the new geological thinking was in terms of hundreds of millions (not billions) of years.

There was enough evidence to support the conclusion that living things had also changed down the ages, but there was nothing solid to suggest how or why it had happened. There were no clues from genetics because nobody knew about genes and scarcely anybody knew about the laws of inheritance. In fact, the Austrian Monk, Gregor Mendel (1822–1884), had spent seven years meticulously revealing those laws by crossing sweet peas in the garden, though he could only guess at the 'elements' that carried the inherited features. Nobody came to help him; indeed the few distributed copies of his published findings failed to arouse any interest during his lifetime. Only after botanists rediscovered the papers in 1900 did Mendel became the 'Father of Genetics.' Since then, generations of students have been required to learn 'Mendel's laws' in school biology lessons.

Other people in different places had wondered about the possibilities of evolutionary change long before 1900, and even before Darwin. A well-known attempt to explain why change would occur was offered by Jean-Baptiste Lamarck (1744–1829), who was then professor of invertebrates at the French National Museum of Natural History. He had actually been appointed professor of the natural history of insects and worms—which indicates the depth of zoological knowledge at the time. It was Lamarck who came up with the word 'invertebrates' and he did much to advance their classification.

Lamarck began to publish his ideas on evolution in 1801, explaining that in response to environmental change, organisms would need to make more or less effort in using their various parts and features. Greater use, he believed, would lead to further tissue development, and less use would lead to atrophy and possibly permanent loss. Somehow, these changes, acquired during the life of an organism, would get passed on to the next generation. It must have seemed quite plausible but nobody could show that any change, acquired during a parent's life, was ever biologically inherited by his or her offspring so Lamarck's explanation of evolution was rejected by influential scientists of the day. In his old age Lamarck went blind and was cared for by his daughter. He died in poverty and was buried for a few years in a rented grave. Nobody knows where his remains now lie.[7]

Since his death, Lamarck has been credited for his work on invertebrates and for recognizing the fact of evolutionary change even though he couldn't explain it. Charles Darwin wrote that 'Lamarck was the first man whose conclusions on the subject excited much attention ... he first did the eminent service of arousing attention to the probability of all change in the organic, as well as in the inorganic world, being the result of law, and not of miraculous interposition.' In a footnote to his tribute to Lamarck, Darwin also remarked 'It is curious how largely my grandfather, Dr Erasmus Darwin, anticipated the views and erroneous grounds of opinion of Lamarck in his *Zoonomia* ... published in 1794.'[8] As far as is known, Erasmus and Lamarck were not aware of each other's work.[9]

By 1858 Charles Darwin had spent more than twenty years collecting and marshalling the supporting evidence for his own explanation of evolution but still he felt he needed to do a lot more. And then, one day in June, he received a letter from the Moluccas in what was then the Dutch East Indies. Darwin was dismayed to find that it contained an explanation of evolution that was virtually the same as his own. The accompanying note politely asked Darwin to read the summary, and see if he thought it sufficiently important to show to Sir Charles Lyell. The letter was from Alfred Russel Wallace, a self-taught British naturalist who was collecting specimens and studying the natural history of the Malay Archipelago. Wallace, a modest man with no family wealth behind him, would achieve exceptional scientific distinction during his life but was not well known at that time.[10]

Darwin, who felt an urgent need of advice, at once contacted Lyell as well as his good friend, J. D. Hooker (an eminent botanist who would become director of the Royal Botanic Gardens, Kew). 'I never saw a more striking coincidence', Darwin wrote to Lyell. 'If Wallace had my M.S. sketch written out in 1842, he could not have made a better short abstract! Even his terms now stand as Heads of my Chapters ... I shall of course at once write & offer to send to any Journal.'[11]

Lyell and Hooker swiftly came up with a plan. Within a fortnight they were reading Wallace's summary to the Linnaean Society, together with an extract from one of Darwin's manuscripts and an outline of evolution that Darwin had once put in a letter to a friend. The presentation was printed in the August issue of the *Journal of the Proceedings of the Linnaean Society of London* with the title: 'On the tendency of species to form varieties; and on the perpetuation of varieties and species by natural means of selection, by C. R. Darwin and A. R. Wallace.' And so it was that in the annals of science the theory of evolution by natural selection was attributed, independently, to Darwin and Wallace—with names in that order.

In everyday conversation, incidentally, the word 'theory' is often used to mean nothing more than a supposition, and that usage has led to a lot of misunderstanding over the years. In the language of science it makes no sense to dismiss the most widely-accepted explanation of something as 'just a theory.'

Darwin cut short the book he was working on and was ready in the following year for publication of *The Origin of Species by Means of Natural Selection.* The initial print run of 1,250 copies sold out on the first day.[12] Six editions of *The Origin* were printed during Darwin's life. As for Wallace, he was more than content. Writing home from the Far East he related how he had 'sent Mr Darwin an essay on a subject upon which he is now writing a great work. He showed it to Dr Hooker and Sir Charles Lyell, who thought so highly of it that they had it read before the Linnaean Society. This insures me the acquaintance of these eminent men on my return home.'[13]

There is nothing to suggest there was ever less than mutual respect and a good relationship between Darwin and Wallace, although it can be argued that Wallace was badly treated.[14] Darwin admits to feeling uncomfortable about the way the Linnaean presentation was handled; remarking, 'I was at first very unwilling to consent, as I thought Mr Wallace might consider my doing so unjustifiable, for I did not then know how generous and noble was his disposition.' Whatever the ethics of the affair, in practice it did Wallace no harm at all. It linked his name with Darwin's, and gave his 'essay' a lot more prominence than it is likely to have achieved otherwise. And Wallace always accepted the older man's priority, referring to their theory as 'Darwinism' even in the title to his own book on evolution, published in 1889.[15]

Wallace records that he spent three evenings writing up the summary that he sent to Darwin but it is wrong to think (as is sometimes written) that the whole outline came to him while he was lying in bed one day with a fever. What did come to mind in that way were the words of Thomas Malthus that Wallace had read some years earlier. It was Malthus' *Principle of Population* that made the penny drop, so to speak. Wallace suddenly realized that the enormous and constant destruction of plants and animals from disease, predation, famine etc. would not be random, it would be selective. He knew first-hand how much individual variation there was in nature and he reasoned that survivors would tend to be those best fitted for survival in each generation. In that way, change would be adaptive.[16] Malthus had given Darwin exactly the same flash of insight back in 1838.[17] One can imagine that after *The Origin* was published scores of people would have been smacking their foreheads and saying 'why didn't I think of that!'

Evolution and society

Darwin barely mentioned human evolution in *The Origin.* All he said on the topic was 'In the distant future ... light will be thrown on the origin of man and his history'.[18] In fact the word 'evolution' does not appear in any context in early editions of *The Origin*, and neither does the term 'survival of the fittest.' They were not Darwin's original terms. But of course people were soon expressing their own opinions about Darwinism in relation to humanity, and it seems probable that much common understanding today is derived not from Darwin's work but from 'social Darwinists' and others who were soon putting their own stamp upon it and applying their different versions to human society—long before science had even begun to reveal the complexities of human heredity.

And the social theorists came in a variety of political stripes. On the left, Karl Marx (1818–1883) was particularly in tune with the 'struggle for existence' and it has long been recorded that he offered to dedicate the English edition of *Das Kapital* to Darwin, although this has been disputed by scholars who have studied the correspondence in question.[19] The Russian anarchist, Peter Kropotkin (1842–1921), believed that communities without state control would naturally come to depend on mutual aid and cooperation because that was the true nature of humans as a social species. At the other end of the political spectrum, the British polymath, Herbert Spencer (1820–1903), who did coin the phrase 'survival of the fittest', also thought the state should have a minimal role but for quite different reasons. In Spencer's logic the state could not be trusted to accept the harsh realities of life. Spencer could accept a measure of compassion but he believed that state intervention was an obstacle to progress if it stood in the way of social evolution by favouring those who contributed least to society.[20]

From notions of social evolution it was easy to move on to the idea that humans should be more selectively bred so as to bring about a general improvement in human quality. The word 'eugenic' (well born) was introduced by Darwin's remarkable half-cousin, Francis Galton (1822–1911), who pioneered the use of fingerprints and statistical methods, and wrote about the technicalities of subjects as diverse as blood transfusions, intelligence and meteorology.[21] So-called positive eugenics would provide economic incentives to encourage good breeding stock to have larger families. Negative eugenics would discourage or prevent inferior individuals, such as the feeble-minded, from breeding at all. In the United Kingdom, incidentally, the term 'feeble-minded' would come to be carefully defined in the Mental Deficiency Act of 1913. The Act was replaced in 1959 by the Mental Health Act.

It is worth emphasizing here how little was known in the nineteenth century—even about reproduction let alone heredity. A Dutchman, Anton van Leeuwenhoek, using home-made versions of the newly-invented microscope, had reported in 1677 seeing sperms swimming in semen, but who could say they were not some sort of parasite? A century or so later, Italian biologists showed that the semen of male frogs was essential to fertilize eggs as they emerged from the females. This experiment involved fitting male frogs with semen-proof trousers before they were allowed to clasp the females. But even that didn't show whether the 'spermatic worms' or only the liquid part of semen was needed.[22] Not until 1854 was it discovered that the sperm cells of frogs enter the egg at fertilization, and it was 1876 before a German biologist, Oscar Hertwig, working with sea urchins, saw that the nucleus of a sperm cell (with visible chromosomes) joins together with that of the egg during fertilization.[23] Such was our understanding of living nature even as the Industrial Revolution was changing the world in which we lived.

In the early 1900s eugenics was supported by prominent and respected people. They included Theodore Roosevelt, Winston Churchill, H. G. Wells, George Bernard Shaw, John Maynard Keynes and many others around the globe. Selective breeding in humans was not a new idea (it dates back to Plato) but Darwin had unwittingly provided a scientific way of looking at it. Eugenics was seen by its supporters as sound and sensible policy for the good of society. This sort of thinking was widely denounced after the atrocities of Nazi Germany and the eugenics movement soon collapsed, but reproductive technology continued to advance. Today, modern genetic screening can offer couples the choice of terminating a pregnancy that would produce a genetically disabled baby. Interventions of this sort, and other genetic manipulations in humans, have been called the 'new eugenics' by some.[24]

In all respects, evolution by natural selection presented a completely new outlook on life. Most people, even those who were not particularly religious, still assumed that a Supreme Being with some sense of purpose must somehow be in control. Now, there were scientists claiming that evolution was a game of numbers that followed blindly and inevitably from the facts of reproductive potential and inherited variation. Evolution was continually playing catch-up with environmental change, nothing could know where it was headed in geological time, and a massive waste of individual lives had always been an integral part of the process. In fact it is even more wasteful than anybody realized; it is now estimated that more than 99 per cent of species that have existed are now extinct.

If this appeared as a bleak revelation, it was made worse by the social theorists who tried to apply it to human society and emphasized the competitive aspect of it. In reality, evolution is a change in the mix of gene types in a population as the more successful combinations leave more offspring and so increase in frequency relative to the others. Long-term reproductive success is what counts in the end and the principle is the same for all forms of life—from mushrooms to mammoths. Winning combinations are always a trade-off between costs and benefits over time. The bird that lays more eggs will not outbreed its rivals if it cannot also raise more hungry chicks. A thorn-bush that spends more of its energy in growing bigger thorns would be losing out if the threat from browsing animals is decreasing. In social animals evolutionary success might include better parenting or possibly cooperation between individuals.[25] Not all games are settled by flooring an opponent, and evolutionary benefit does not only result from the immediate outcome of competitions over one thing at a time. Life is a lot more complicated than that.

Prior to Darwin an interest in natural history was entirely appropriate for country parsons because it had long stood to reason that the wonders of nature were evidence of the Creator's powers. In the late seventeenth century the English naturalist, John Ray (1670–1705), had written volumes on that way of thinking (known as natural theology), and the subsequent work of William Paley remained highly influential throughout the nineteenth century.[26]

So, not surprisingly, opposition to Darwinism was particularly vigorous on religious grounds, but by the end of the 19th century, when the idea of evolution had been more broadly accepted by the public, the scientific community had come to doubt whether natural selection really could explain it. After the rediscovery of Mendel's data, leading scientists were divided on the question of whether his findings could account for all the variability to be seen in different species. But eventually, after decades of work by geneticists and statisticians, and with more information from the fossil record, a depth of understanding was reached that was generally acceptable to scientists in all fields. It was summarized in 1942 by Julian Huxley in his book *Evolution: the Modern Synthesis*.[27] After that, Darwinism became Neo-Darwinism.

Evolutionary biologists today attempt to resolve remaining uncertainties such as the relative importance of large versus small mutations, the action of genes in developing embryos, and the evolutionary importance of epigenetics, which is the chemical tagging of genes in sperms and eggs that affects the way they work after fertilization. These are important issues but they are peripheral to the evolutionary principles that really do make up, as is often said, 'the cornerstone of modern biology'.

11 Post-industrial economics and ecology

The social and economic changes that resulted from industry were spread throughout the world, directly and indirectly, by colonization. Imperialism is a rich seam for historians of all political persuasions and no quick and simple overview can encompass the diversity of case histories. Nor do the ex-colonies have much in common today as they include most countries of the modern world. Ethiopia and Nepal, countries in which I have spent some years, are among the few in Africa and Asia that were never fully colonized and they rank among the world's poorest nations.[1]

Economics

What can be said, indisputably, is that colonization by European powers created new markets far away from Europe, and a massive expansion of global trade for raw materials and manufactured goods. Technology and policy worked in tandem to achieve this. Steam ships and railways lowered transport costs dramatically, while telegraph and telephone worked wonders with communications. It is estimated that in the late 19th century the cost of transport between America and Europe fell from about 80 per cent of the price of transported commodities to less than 20 per cent. Before the nineteenth century, transport costs for most commodities exceeded the price of goods in the country of origin.[2]

It is also true that as industries contributed more to the growth of economies, so the incidence and impact of what economists call 'externalities' also grew. Externalities can be thought of as spillover, to third parties, of costs or benefits resulting from economic transactions. If the spillovers are not built in to the economics of the transaction then they are externalities. Negative externalities include pollution in its many forms but externalities can also be positive. Research and development, for instance, might be funded for a specific purpose but also have unforseen benefits to the wider community.

As capital markets became more integrated, Great Britain became the world's banker and many European countries followed Britain's example in reducing or abolishing tariffs the better to share in the burgeoning world trade. International free trade is said to have reached a peak in the third quarter of the 19th century, by which time both Germany and Italy had become unified states. Some protectionism was revived after 1880, though foreign trade continued to increase. Available statistics indicate that by 1913 the volume of foreign trade per capita had grown to more than 25 times what it had been in 1800.[3] Real wages and living standards rose globally but grew much more rapidly in the industrialized countries—as did the social middle class. Until the early eighteenth century, living standards between different geographical regions were probably not strikingly different but in economic terms the world had clearly divided in two by the end of the nineteenth century.[4]

As for economic thinking, the doctrines of what is now known as the Classical School, beginning with the writings of Smith, Malthus and Ricardo, were amply tested by all these developments and they were expanded and systematized by others who followed. In the English-speaking world John Stuart Mill (1806–1873) is perhaps the most famous of those who lived entirely within the 19th century and he is still much quoted today, though not so much for his economic insights as for his pithy assertions about society and life in general. An online search for the quotes of J. S. Mill is likely to turn up dozens of them.

John Stuart was a child prodigy, a phenomenon quickly recognized and exploited by his father, James Mill, who educated the boy intensively with the aim of producing an intellect best able to further the cause of Utilitarianism. I don't want to stray here into the minefield of moral philosophy, but the fact is that economics had its origins in political economy which in turn grew from moral philosophy. Developments in the nineteenth century raised new concerns about rights and wrongs in regard to wealth and welfare and they challenged some of the best brains of the time.

Utilitarianism was not the brainchild of James Mill alone. Its first formulation is usually attributed to his friend Jeremy Bentham (1748–1832) and some elements of the doctrine were not new at all. The nub of the thinking was that the morally right action is that which has the highest utility in providing the most good for the greatest number of people. 'Good' in this sense can be equated with pleasure or satisfaction. This might seem to be a pretty simple notion until you try to apply it to all the complexities of an industrial society. Bentham, for instance, considered all forms of happiness to be of equal worth but J. S. Mill thought otherwise. For him, intellectual pleasures were of a higher order than the sort of sensual pleasures that we share with other animals.[5]

J. S. Mill set out his own version of Utilitarianism in 1863 in a book of about one hundred pages.[6] This was fifteen years after the first publication of his most widely-read work, *Principles of Political Economy*, which ran to seven editions and remained a standard text well into the twentieth century.[7] It is said that J. S. Mill was the last of the great political economists to champion the market system, but the extent to which he was ever in favour of laissez-faire is still debated.[8] No matter how he may be judged, it is obvious that his economic thinking was influenced by his utilitarian principles, and it is equally clear that concerns about social issues, and what might be seen as socialist leanings, featured more prominently in his later works.

In America, the country that was forging ahead to become the world's leading industrial nation, socialist thinking was slower to make its mark. The British notion of worker solidarity held less appeal to tens of millions of migrants with a frontier mentality, though the more industrialized of the States did begin to enact labour laws toward the end of the 19th century. In general, the progress of labour legislation outside western Europe, Australia, and New Zealand was slow until after World War I.[9]

It would be mere coincidence if the turn of a century happened to mark some defining period in history and in this instance the momentum of the nineteenth century continued until the First World War.[10] By that time, Britain no longer ruled the waves but about 85 per cent of the earth's surface was, or had been, under British or other European rule and there were spheres of influence and commercial treaties that further extended the economic influence of major powers to cover nearly the entire globe.[11]

People and the wild

The word 'ecology' was coined a few years after Darwin's *Origin* was published but naturalists were already familiar with the idea of interactions in nature, including the concept of closely-packed niches. Darwin was personally quite taken with the thought that the abundance of red clover and certain other wild flowers might be influenced by domestic cats. It was reasoned that populations of bees, which pollinated the flowers, were suppressed by field-mice, which destroyed the honeycomb nests, and the mice were preyed upon by cats. Bees' nests, it appeared, were more plentiful where people kept more cats.[12] But, generally speaking, the study of food webs in the wild would probably have been regarded as a very trivial pursuit for townsfolk of the nineteenth century.

In 1800 the human population, by best estimates, was about 0.97 billion. Not until 1910 did the United Nations and U.S. Census Bureau began to collect population data, after which it could be seen that the global population had more than doubled to reach 2.07 billion in 1930. There were huge regional differences. Within the nineteenth century Africa's population increased by about 24 per cent and that of Asia by 49 per cent. Over the same period the population of North America grew from about seven million to 82 million—an increase of over a thousand per cent.[13] Even without new industries an extra billion people would have needed proportionally more farmland, fuel, housing, clothing and so on, largely at the expense of wild nature.

The concept of the ecological footprint was developed quite recently but it can be applied to any kind of society from any time if enough information is available. By averaging the productivity of different types of biologically productive land, such as cropland and pasture, one hectare (or acre) of land can be used as a global accounting unit. On this basis it has been estimated that the ecological footprint per person in 1820 was about 1.1 global hectares.[14] For what it is worth, this compares quite closely with the 1.48 local hectares per person calculated for an 'eco-village' in Hungary, where a community lives without mains electricity and makes every effort to have minimal impact on nature.[15] (but see chapter 15 for a critique of footprint accounting).

In industrialized regions there was also pollution to contend with. Filth and squalor did not arrive with the Industrial Revolution so to some extent it was only more of the same as towns grew in size. But when factories as well as households were belching smoke from coal fires, and with industrial chemicals flowing into rivers and lakes, pollution in the late nineteenth century had no precedent and it was not brought under control until well into the twentieth century. In towns, smoke was trapped in the streets by fog to form a sulphurous, greenish 'smog', commonly known as a 'pea souper', that was frequently lethal to those most vulnerable. On occasions in England, cars and buses would be guided through London, Manchester or Birmingham by somebody walking a few steps ahead and waving a torch. Pedestrians had serious difficulty in finding their own front doors. Thousands were killed by the last London smog, a particularly bad one, in 1952.[16]

Similar air pollution occurred in cities elsewhere. In Donora, Pennsylvania, thousands were made ill by air pollution in 1948, and the event led to the first federal legislation to control air pollution in the USA.[17] While smogs were a phenomenon of big towns, water pollution spread widely so that entire aquatic ecosystems were destroyed as rivers and lakes became toxic. With today's legislation, grossly polluted waterways are far fewer, at least in western countries.

On the plus side, nineteenth century Europe reached a new level of freedom of thought in comparison with the preceding few centuries. Jeremy Bentham, for example, had become a well-respected figure by the 1820s, despite being an atheist and ahead of his time in advocating votes for women and the legalizing of homosexuality. Bentham, it was, who donated his body to science and helped to stop grave-robbing and murder by supporting the Anatomy Act of 1832. The Act made more corpses lawfully accessible to the medical profession. After dissection of Bentham's own body, in accordance with his will, his skeleton was padded out, dressed in his personal clothes, and displayed sitting in a cabinet. This strange 'auto-icon' is still on view at University College London.[18]

Notwithstanding all this enlightenment, the prevailing western view of the natural environment was still the biblical one: nature was there to be tamed and mastered because man should have dominion over every living thing. Religions throughout the world combine the sacred with utility in one way or another. Sacred groves have been maintained for thousands of years and, at least in the past, their carers have successfully combined strict preservation with some sort of sustainable use.[19] The appeal of domination is easy to understand in the minds of colonists, fighting their way through unfamiliar and dangerous territory. And it is not so surprising among those who were comfortably at home and were well aware that vast tracts of wilderness still existed around the world. Worries about adverse impacts of human settlement, growth and progress would have been very low on the list of concerns for most people.

But there are always exceptional individuals and since antiquity there have been those who have influenced others with a love of wild nature. Henry David Thoreau (1817–1862), spent two years living in a cabin in woodland near Walden Pond, a lake in Massachusetts, USA. His book, *Walden*, became a classic. John Muir (1838–1914) was born in Scotland but in 1892 he founded the Sierra Club, an American conservation organization which now claims to have 2.4 million members and supporters. Some sporting shooters were also exceptional achievers in their conservation efforts. Theodore Roosevelt was one of them. As American President (1901–1909) he was able to have scores of areas placed under various categories of public protection, including national forests, game preserves and bird reservations. Yellowstone, in the USA, had already become the world's first national park in 1872.

Roosevelt also helped to save the American bison from extinction. Popularly called buffalo, these animals had been shot in their millions to the point where only a few hundred remained in the wild. But not even the American President could save the passenger pigeon, the numbers of which were said to darken the sky as they passed over in vast flocks. The last specimen died in Cincinnati Zoo in 1914. Passenger pigeons had been killed because they were a ready source of meat but, strange as it now seems, the first conservation organizations with an international agenda were set up in response to the huge commercial trade in feathers.

Around the turn of the century it was fashionable for women to wear feathered hats and trade was booming. Birds with attractive feathers were being killed in hundreds of millions, mainly at nesting time when males were in breeding plumage. London was the hub of the international trade.[20] Some people, especially women, were outraged by this destruction and came together to put a stop to it. The Society for the Protection of Birds was formed in England in 1889, and the first Audubon Society (Massachusetts) was set up in 1896. The former became the Royal Society for the Protection of Birds (1904), and the National Association of Audubon Societies was incorporated in the USA in 1905.

As for the rest of the world, local hunting and gathering was expanding along with agriculture, but the impact of European sport-hunters had also been substantial in the colonies. Recognizing this, the colonial powers in Africa at the time (Great Britain, France, Italy, Portugal and Spain) agreed in 1900 on the first treaty for the protection of wild animals in Africa to 'ensure the conservation of diverse wild animal species ... which are useful to man or inoffensive'. The treaty was known as the London Convention and although it may have been useful as a stepping-stone to later legislation, it never came into force.

12 Living it up and crashing down

Considering that the first half of the twentieth century was dominated by two world wars and the worst economic depression in history, the American way of life in the 1920s may, on reflection, seem surprising. By the end of the 1920s, most homes in American towns had electricity, and the householders were intent on making the most of it—along with everything else that made up the good times of the 'roaring twenties'. As one historian describes it: 'By 1929, American families spent over 20 percent of their household earnings on such items as phonographs, factory-made furniture, radios, electric appliances, automobiles, and entertainment. What people couldn't afford, they borrowed.'[1]

There is no obvious explanation for the roaring twenties but war certainly had a lot to do with it, and to explain the war, even in the simplest of terms, we must go back a little further in time. It is generally acknowledged by historians that war, depression, and more war followed each other in the twentieth century as consequences of power struggles that had started in Europe several decades earlier. The unifications of Germany and Italy, in particular, caused major stresses in Europe's already complicated balance of power, drawing the continent into a chequer-board of shifting and uneasy alliances between anxious nations.

War and depression

By 1907, Britain, Russia and France made up the Triple Entente (basis of the Allies), while Germany, Austria-Hungary, and Italy constituted the Triple Alliance, but the two alliance systems were anything but stable as they built up their military arsenals and plotted and manoeuvred with increasing paranoia. Hostilities in the small Balkan states were a significant worry to Austria and Russia because possible outcomes could favour one side or the other. The flashpoint came in 1914 when a Serbian nationalist assassinated Austria's Archduke Ferdinand and his wife in an attempt to further Serbian territorial claims. Austria retaliated against Serbia, Russia took Serbia's side, and so began the progression to the Great War that was supposed to end all wars.[2] Italy joined the Allies, but the Ottoman Empire (centred on Turkey) and Bulgaria joined the Triple Alliance to form the Central Powers.[3] The war dragged on for four years and caused total casualties in the tens of millions.

Because much of the world was controlled or influenced by a few European nations the Great War quickly became a global rather than regional crisis, but nobody expected it to last as long as it did, or to have such devastating consequences in terms of loss of life and human suffering. Earlier wars had never been so mechanized and the slaughter and maiming was truly industrial in scale.

In geopolitical terms the Great War caused the fall of four empires—those of Germany, Russia, Austria-Hungary, and Turkey. The last imperial dynasty of China (Qing, also called Manchu) had collapsed in 1912 partly because of growing European pressures around its borders. After the Great War there was also a marked rise in, or strengthening of, independence movements in British and other colonies. The movements gained traction from the fact that belief in the natural superiority of European leadership had become a lot harder to uphold.

Japan's role in the war was limited but significant. After capturing the German port of Tsingtau on the Chinese mainland, the Japanese navy kept the regional sea lanes open for allied shipping. As a consequence, Japan increased its influence in China, lifted its status as a global power, and boosted its economy by supplying the Allies with war goods. It is reported, however, that modern Japan exhibits little memory of the nation's role in the Great War.[4]

America, on the other hand, emerged from the conflict with lasting economic and political benefits. To begin with, America tried to maintain trade with both sides but the actions of German submarines (U-Boats) in sinking allied supply ships became intolerable in a situation where passenger vessels could be suspected of being armed and carrying munitions. American opinion began to firm up in 1915 after the sinking of the British liner *Lusitania* with the loss of more than a thousand lives—including 128 Americans. The subsequent sinking of American ships persuaded President Wilson to break off diplomatic relations with Germany and enter the war on the side of the Allies in April 1917.

Before the outbreak of war, America had been building up its industrial capacity and so was well prepared to produce the necessary materials and supplies. The army and navy contracted work to the private sector, which resulted in an enormous increase in federal spending, and also the creation of new agencies for controlling specific areas of the economy, such as food, fuel, railroads, and many others, which traditionally would have been left to the market to sort out. The government raised money mainly through a combination of higher taxes and the sale of war bonds (Liberty Bonds).

The Americans surprised everyone (not least the Germans) with the speed with which they were able to mobilize. Conscription was introduced, and by May 1918 more than a million US troops were stationed in France.[5] An armistice with Germany was signed on 11 November 1918, and peace terms were imposed on Germany with the signing of the Treaty of Versailles the following June. Another outcome of that treaty was the formation of the League of Nations, an organization intended to promote international cooperation and maintain peace and security.

The war improved America's economy, lifting the nation out of debt and building credit to the tune of $6.4 billion by 1919. But in the opinion of economic historian, Hugh Rockoff, the economic and administrative lessons learned by Americans during the war years may have been even more valuable than the immediate prosperity because similar thinking would be needed again during the Great Depression.[6] Of course, not all economists would agree with Professor Rockoff any more than they would all agree about why an economic downturn sank America into a Great Depression.

With all the confidence of the roaring twenties, investors who were keen to have a stake in the booming stock market would borrow two thirds or more of the face-value of securities (stocks) to buy them 'on margin'. In this arrangement the investor usually borrows from a broker, using securities as collateral for the loan, and the loan must be repaid with interest. As a safeguard, the broker sets a margin in the deal so that in the event that the securities fall in value below a certain point, the investor must pay off some of the loan to restore the margin—otherwise the broker will restore it by selling off some of the investor's securities. Published reports differ somewhat, but according to one respected source more than $8.5 billion was out on loan in this way by August 1929.[7]

There were some price falls in early October that year but nothing drastic happened until panic set in on October 24th when prices crashed in a massive sell-off. Leading bankers did what they could to restore confidence and stability but panic is highly contagious and the recovery was short lived. October 28th will be forever known on Wall Street as 'Black Tuesday', the day when the market crashed again and could not be revived. The downturn was global; industrial nations around the world were struggling with declining trade and productivity and were technically in recession by 1929. Some countries, including Japan, got off relatively lightly and soon recovered, but in America the downward trend became a dive.

I should explain here that a depression is now commonly defined as a particularly severe recession that lasts for two or more years, and a recession is usually declared when the gross domestic product (GDP) has declined for two successive quarters. At the time of the Great Depression there was no collection of economic data with which to calculate a GDP or any other national index of economic performance, but somebody was working on it.

Simon Kuznets (1901–1985) was born in Russia to Jewish parents. He had partially completed degree studies in economics by 1922, in which year he migrated to America and in quick succession gained a B.Sc., M.A. and Ph.D from Columbia University. In 1927 he joined the research staff of the National Bureau of Economic Research (NBER) and developed a system of national accounting that would eventually become the basis of the global standard. More immediately, and at a crucial time, Kuznets and his team provided the US Senate with the sort of data, accounting procedures and methods of analysis that were needed for improved economic planning.[8]

And so the concept of GDP was launched. Basically, it means the value of all goods and services produced in a nation over a given time. It is a measure of economic activity but does not measure human welfare. Kuznets never claimed otherwise.

The social impact of the Great Depression was far-reaching. Black Tuesday has given rise to urban myths and flippant stories about mass suicides that never occurred but the stark facts are well documented. In the years that followed the market crash, small banks failed in their thousands (there were too many of them[9]), businesses closed, families lost their homes or farms and tenant farmers were evicted. In 1932 and 1933 nearly a quarter of the work force was unemployed and destitute people lived in shanties and queued in the streets for bread and soup. Recovery did not begin until 1933. Output then grew each year until 1937, after which there was another downturn.[10]

In earlier times, silver and gold coins had served three functions: they were a handy medium of exchange, they were good for storing wealth in secure places, and they were worth what they were supposed to be worth because silver or gold is its own measure of value. Early paper money also promised to pay its declared value in silver or gold, and for a few decades before 1914 the currencies of most industrial nations were similarly tied to the value of gold. This determined exchange rates and helped to keep currencies stable because each country on 'the gold standard' was not supposed to print more money than was backed by gold reserves—which were constantly being shuttled between the nations involved. The outbreak of war put an end to the gold standard; alliances changed and nations needed money regardless of stability and exchange rates.

The gold standard was tried again after the war but it didn't survive the Great Depression because linked economies were all being dragged down together. Nations came off the gold standard again and tried to pull out of the depression in their own ways. More often than not, that meant trying to protect national economies by imposing tariffs on imports and cutting back on spending—which reduced international trade and made matters worse according to many analysts.[11] The Federal Reserve, America's central bank, was set up in 1913 so was still young. Some say it didn't do enough; others say it did too much, or acted too late. Milton Friedman, the twentieth century's best-known advocate of the free market, said plainly that it did the wrong things. Moreover, Ben Bernanke agrees with him, and Bernanke was chairman of the Federal Reserve from 2006 to 2014.[12]

As for government interventions, for much of the roaring twenties Calvin Coolidge had been president and he was famous for intervening, or even speaking, as little as possible. Indeed, when his death was reported in 1933, one well-known wit promptly asked 'how could they tell?'[13] Coolidge was succeeded by Herbert Hoover, who held office from 1929–1933. Hoover worked hard at strategies that were intended to get the economy moving again, but he stopped short of federal controls on business, or directly trying to manipulate prices or the value of the currency. He lost the election to Franklin D. Roosevelt—the only American president to be elected for four terms.

Roosevelt came to office promising a 'New Deal' for ordinary Americans. Using Kuznets' national accounting for budget planning, and beginning with the urgent needs of the unemployed, the New Deal programme went on to extend federal control into areas of finance, banking, business and agriculture. It included, for example, federal insurance for bank deposits and cash subsidies to farmers. By 1939 Roosevelt had introduced social security benefits for the aged, widowed, and unemployed, and, in some industries, maximum working hours and minimum wages. Still, the American emphasis on personal responsibility ensured that benefits were to be based on worker contributions and could not be expected as handouts—which was the American perception of some of the existing European schemes.[14] Roosevelt's New Deal programmes were largely followed by his successor (Harry S. Truman) and were eventually accepted by both sides of American politics.[15]

In Europe, the ravages and outcomes of the war affected nations very differently. Much of the continent was in debt to

America and was still being supported by American credit, so the depression for some European countries was made worse by the Wall Street crash and the subsequent credit drought. Germany was in a particularly poor state because, under the Treaty of Versailles, Germany was billed with reparations that she was obviously unable to pay. Things were so bad in 1931 that President Hoover, fearing the knock-on effects of a total German collapse, proposed a one-year moratorium on Germany's repayments. It was vigorously opposed by France but came into effect later that year. The moratorium deprived France and Britain of a precious source of funds which they had been using to repay America, and it failed to halt the economic deterioration in Germany.[16]

And worse to come

Poverty was not the only problem in Europe, and the depression was not the only cause of it. Following the Russian revolution in 1917, the Soviets, under Joseph Stalin, were discovering the hard way that while Marx may have been an impressive economic theorist, he had little understanding of human attachment to land, or the wider importance of family, relatives and small groups in our make-up as social primates. Farmers the world over are known to slave devotedly on their family farms, but Soviet peasants worked with predictably less enthusiasm when family farms were forcibly replaced by state collectives. This contributed to a severe famine in Russia during 1932–33.

In the Soviet far east, Stalin controlled Mongolia as a puppet state in the hands of local warlords. East of that, the region known as Manchuria was a similar puppet state of Japan. Japan has few natural resources and Manchukuo (as the Japanese called their colony) had coal deposits and farmland that would support further Japanese expansion into China. The border had long been disputed by the locals, and in 1939 Russia and Japan went to war over it. In a battle that involved over 100,000 troops, the Japanese were comprehensively defeated—an outcome that may well have influenced Japanese decision-making a couple of years later.[17]

In Italy, Benito Mussolini became popular after his appointment as prime minister of the Kingdom of Italy in 1922. But after a few years he was running the country as fascist dictator with a yearning to rule over a bigger empire. In the Horn of Africa he already possessed the colonies of Eritrea and Italian Somaliland, both of them bordering Abyssinia (now Ethiopia), which Mussolini invaded and occupied in 1935. It was liberated in 1941.

Mussolini might have held Abyssinia for a lot longer than he did, had it not been for the ambitions of Germany's leader, Adolf Hitler. Upon leaving the army in 1919, Hitler joined a workers' political party in Munich that became the National Socialist German Workers' Party—the Nazi-Sozis or Nazis as they became universally known. It was not a smooth path to the top, but Hitler became party leader, and in 1933 President Hindenberg made him the chancellor of Germany. Within a matter of months, Hitler had absolute power and all non-Nazi parties and even labour unions had been outlawed.

Under the Nazis, employment and the economy grew on the basis of huge public works projects and a rearmament programme that was in breach of the terms of the peace treaty. From the start, Hitler was intent on reuniting all German people within a single state, and expanding their territory to the east. This was Hitler's goal and he pursued it with a passion, showing 'astonishing skill in judging the mood of the democratic leaders and exploiting their weaknesses' as he made his moves.[18]

Hitler found a ready ally in Mussolini, who by then was having to cope with sanctions imposed by Britain and France in response to Italy's invasion of Abyssinia. In 1938, Germany annexed Austria with virtually no resistance, after which Hitler extended German control over Czechoslovakia, and recovered some territory from Lithuania that had been detached from Germany by the Treaty of Versailles. He then amazed everyone by managing to secure a non-aggression pact with Stalin so that he could invade Poland and give the Soviets a slice of it. Only when Germany launched a blitzkrieg against Poland in 1939 did Britain and France declare war. France and Britain thereby became the principal allies against Germany and Italy—the Axis powers. But the Nazis, within a few months and with the same overwhelming invasion strategies, went on to occupy Denmark, Norway, Belgium, the Netherlands, Luxembourg and most of France. Britain held out against Germany and famously defeated the Luftwaffe in the Battle of Britain.

In 1941, encouraged by having swept so easily through much of Europe, and in disregard of the non-aggression pact with Stalin, German forces invaded the vast, freezing lands of the Soviet Union. They came close to another conquest but were eventually repelled—with the Red Army in pursuit. In that same year, Japan plunged into the war.

Japan's exploits in China and, by then, in French Indo-China, had provoked America into imposing oil and other embargoes on Japan. The Japanese, having quite recently been beaten by the Soviets, might have been wary of joining the Germans against Stalin, but Germany's swift military successes in Europe would seem to favour a golden opportunity for Japanese expansion elsewhere. The Dutch East Indies, in particular, had abundant resources and the Netherlands was scarcely in a position to protect them. The worry for Japan was the presence of the American naval base at Pearl Harbour in Hawaii, so on 7 December, 1941, Japan launched its surprise air attack.

From Japan's point of view, the attack on Pearl Harbour was a great success, causing devastation and the loss of more than 2,000 American lives. But it did not destroy the aircraft carriers, and in the longer term it was a disaster for Japan: America declared war the following day. Japan quickly went on to gain control of the Philippines, Dutch East Indies, much of New Guinea, the Solomon Islands and other strategic parts of the western Pacific, thereby ensuring that the horrors of the Second World War would extend across the globe. By the time the Japanese surrendered in 1945, cities and towns from Coventry to Tokyo lay in ruins and the war had claimed an estimated 60–80 million lives, possibly three times more than were killed by the Great War, which by then was more appropriately referred to as WWI. The League of Nations, having proved to be totally ineffective in its role, was replaced in 1945 by a new organization of 51 Member States, calling themselves the United Nations.

Part three: Into the growth trap

13 Starting afresh

In human affairs, the end of the Second World War was a turning-point of such significance that it has reasonably been called 'Year Zero'.[1] Throughout Europe and much of Asia, rubble and wreckage was all that remained of once-busy ports, railways and bridges. From shattered communities the displaced people, refugees and orphans slowly moved through the devastation in their millions. In Western Europe the occupying forces had provisioned themselves from local supplies so there were no food reserves to be found, and famine relief from America was hard to distribute with only the remnants of a transport system.

Postwar world

Once again, America prospered at a relatively safe distance, although more than 400,000 Americans lost their lives in WWII.[2] Historians may argue about why the Great Depression lasted as long as it did, but they are agreed that the war finally brought it to an end. In 1939, before the US entered the war, unemployment was around 15 per cent but when millions of Americans were enlisted in the armed forces, or working on government projects for the war effort, America fast became a land of low unemployment and rising wages. Moreover, under wartime rationing, and with industries focused on military needs, there was limited opportunity for family extravagance so debts were paid down and savings accumulated. After the war, with rationing gone, the housing market back in business, and manufacturers turning out cars and household appliances again, America's economic boom was assured.[3] As Paul Krugman put it, 'the enormous public works project known as World War II ... finally provided a fiscal stimulus adequate to the economy's needs.'[4]

And it wasn't just an economic pick-up. To meet the urgent needs of the war, America had come to depend on close collaboration between federal government, private enterprise (including high-tech universities) and organized labour—thereby improving the conditions for further economic growth.[5] Lessons had also been learned from the aftermath of WWI. This time, in order to restore international trade, it made economic sense to help in the recovery of shattered nations, including Germany, Italy and Japan. The Americans and the British were thinking in these terms even as war raged. As for Britain's own position, it was obvious that there would be no funds available for repaying war debts to America so some US aid was provided under a 'lend-lease' arrangement in exchange for long-term use of British air and naval bases in the western hemisphere.[6]

In 1944, delegates from 44 nations came together at Bretton Woods, New Hampshire, to work out a plan for speeding up economic progress and fostering political stability and peace. Following this meeting, the International Monetary Fund (IMF) was set up with the overall aim of furthering international trade. Among other things, the Fund had the delicate task of trying to stabilize exchange rates without seeming to threaten national independence. Another institution, the International Bank for Reconstruction and Development (now known as the World Bank) was set up for the purpose of financing postwar reconstruction. Inevitably, American and British progress in developing systems of national accounting served to inform much of the discussion at Bretton Woods. This entrenched the use of GDP as a universal index of economic growth and progress.[7]

The General Agreement on Tariffs and Trade (GATT) came

into effect in 1947 with the aim of reducing tariffs and promoting fair trade between all contracting nations. In other words, facilitating globalization. In principle, the aim of expanding trade throughout the world is centuries old, but like so much else, it had become a lot more complicated. The GATT covered tens of thousands of tariff concessions and billions of dollars in trade over a period of 47 years, yet it was only ever intended to be a provisional agreement and was eventually replaced by the World Trade Organization in 1995.[8]

Despite all these efforts, growth in Europe remained slow, and through American eyes that was doubly worrying because an economically weak Europe could be vulnerable to communist expansion. Beginning in 1948, America funded a European Recovery Programme for western European nations. It became known as the Marshall Plan, having been proposed by the US Secretary of State, George Marshall. The aid consisted largely of commodities and materials to support industries. It didn't amount to much as a proportion of GDP of recipient nations, but combined with widespread currency devaluations in Europe during 1949, it was enough to make a big difference.[9]

Stalin refused Marshall aid, and under his system of repression and total command, the Soviet Union forged ahead. With a revolution as well as two world wars in its immediate past there was a huge backlog of 'unrealized growth potential'. The workers, having survived the Nazis as well as Stalin's earlier purges, must have felt they had proved their loyalty and had every reason to hope for a better future.[10] It would be forty years before Party Secretary Gorbachev brought about a relaxation of the command economy, by which time Soviet citizens had grown accustomed to a poorly-performing agricultural sector, and a perpetual shortage of everyday needs.

Not surprisingly, Stalin's wartime alliance with the West didn't last long, though the two ideologies survived side-by-side for a lot longer than Hitler had hoped and predicted. Following the Nazi surrender, Germany was divided into four zones, each administered by one of the four main allies. The American and British zones were later merged, and in 1949, with tensions rising between Soviet and western powers, the French zone was included in the merger to create the Federal Republic of Germany (West Germany). The Soviets then established the German Democratic Republic (East Germany).

Germany and Japan

West Germany achieved what appeared to be a postwar economic miracle and it was not a consequence of Marshall aid, though West Germany was a recipient. It was a result of accepting the insights and adopting the policies of Ludwig Erhard, who would become minister of economic affairs, and then from 1961–1963, chancellor of West Germany. At the time, his proposals seemed counter-intuitive to observers but in 1948 Erhard correctly predicted how the defeated Germans would respond to drastic currency reforms and the abolition of price controls. Erhard's strategies, and the German people, immediately started to work.[11]

For different reasons, Japan's recovery from the war was also phenomenal. Stripped of its pre-war colonies, as well as its wartime conquests, Japan was again short of natural resources so was dependent on imports for its industrial needs. These had to be paid for, and despite the many reforms imposed during the American occupation, and the fact that Japan no longer had defence expenses, the economy was still weak by 1950. In that year, events elsewhere suddenly solved Japan's problems.

The Korean peninsula, liberated from Japan by the Allies, ended up like Germany—split in half by ideology. Korea had been annexed to Japan since 1910 and had, so to speak, no rightful owner waiting to claim it. The recently-formed United Nations found it impossible to create a unified Korea so America and the Soviet Union agreed on an administrative division at the 38th parallel (latitude 38° N). The Soviets were to administer the northern part while the Americans controlled the southern region. It was not a stable arrangement. In 1950, the North, backed by the Soviet Union and supported by China, (by then also communist) invaded the South. The UN went to the aid of South Korea and the ensuing war lasted for three years.

The Korean war, by some estimates, claimed more than two million lives and settled nothing. On opposite sides of a demilitarized zone the two Koreas are technically still at war. But during the fighting, Japan became the main supply depot for UN forces so dollars and materials flooded into the country. It was an economic windfall and the Japanese made the most of it.[12] As in Germany, there may also have been a cultural dimension to the Japanese success. According to some analysts, workers felt a special loyalty to companies, now including household names in the auto and electronics industries, that offered jobs for life and were structured in ways that projected the appeal of a big family or religious institution.[13]

Lift-off and growth

Directly and indirectly, war accelerated social and economic change around the world. People had come to hope for new directions and had turned to different leaders and independence movements. Wartime needs had created opportunities for women to prove themselves capable of doing all manner of men's jobs. Not for the first time, governments had been forced to support medical and technological research with more urgency.

In the history of military conflicts it was usual for more people to die from disease than from battle. In the Crimean War (1853–1856), typhoid alone caused higher mortality than did the fighting. From a medical point of view the First World War has been described as a giant field trial for the administration and sanitation of vast armies. Purified water, better hygiene, typhoid vaccinations and anti-tetanus injections saved countless lives. It was inevitable that modern weaponry would cause a higher proportion of fatalities, but the remarkable fact is that as a result of surgical advances and improved battlefield procedures, the proportion of deaths from wounding in WWI was probably less than half that of the Crimean conflict.[14] The Second World War again focused research efforts in medical and other areas of high priority, and led to government funding and industry collaboration on a scale never seen before.

Many, if not most, of the technical advances were not entirely new discoveries so much as huge improvements on existing ideas, which led to new applications. The spin-off from all the technological effort of the war years fed into peacetime research and development to become the stuff of our global economy. The Manhattan project produced the atom bomb, and now nuclear power produces eleven per cent of all the world's electricity. The plastics industry came of age when common metals suddenly became precious war materials.[15] Radio navigation, jet engines, pressurized cabins, and even crash helmets and seat belts resulted from aviation research. In medicine, blood transfusions were perfected and the first of the antibiotics was mass-produced. The catastrophe of WWII ushered in the modern world, a world upon which the impact of human activity has reached a level that would have been unimaginable a century earlier.

Economies grow along with populations, and populations can grow exponentially for limited periods. It still surprises a lot of people to learn that if you start with one dollar and double your money every day, you will be a millionaire in three weeks. Such is the power of exponential growth. The first few doublings don't grab your attention but the J-curve on a graph takes a very steep climb as the numbers grow. If you really had those doubling dollars for three weeks, after just one more day you'd be a millionaire twice over.

It took more than two hundred thousand years for the human population to grow to one billion (a thousand millions). The number was first reached shortly after the year 1800. But by the time of the Great Depression the total had reached two billion—a doubling time of less than 130 years. By 1974 there were four billion of us so population doubling had taken less than 50 years. The same thing is set to happen again because we added the last billion during a twelve-year period and on current projections there will be eight billion people in the world in only five years' time (2023). Counting eight billion by the second-ticks on a clock would take more than 253 years.

The annual rate of population growth has fallen from its peak of slightly over two per cent in the 1960s to its present rate of 1.18 percent. But the smaller percentage now applies to a much bigger base so we are actually adding more people over a short period than we were in the 1960s—currently some 80 million a year.[16] The reasons for the rapid increase in numbers are easy to understand. The original condition of high fertility and high infant mortality was ended when the main killer diseases were brought under control, but birth control has lagged well behind death control so numbers have been climbing phenomenally for decades. At the same time, modern medicine has prolonged the lives of adults so that a much higher proportion of us live to reach old age. We'll return to that topic in chapter 17.

The surge in human numbers was a boon for the economy but economic growth went far beyond it; real GDP in the top 16 OECD countries rose by an average annual rate of 4.8 per cent during the years 1950–1973. In developing countries, income per person increased by an average of 3 per cent a year over the same period. The growth rate in trade was double that of any earlier time. It was a period of economic growth that has never been equalled and it has gone down in the history of economics as a golden age. Living standards soared along with the demand for workers in the manufacture and sale of refrigerators, TV sets, central heating systems, cars, and all else that was popular under the heading of durable consumer goods. The golden age began with growth rates increasing more rapidly in some countries than in others (fastest in Japan) and it came to an end with falling labour productivity becoming apparent in some countries but not in others before 1973.[17,18]

It was in October of that year, however, that Egypt and Syria launched their attack on Israel and started what would become known as the Yom Kippur War. The Arab oil producers then imposed an oil embargo on those nations that were seen to be supportive of Israel. Non-Arab producers at that time could not make up the shortfall and the price of crude oil quadrupled from US$3 to US$12. Countries that were not specifically targeted were affected nonetheless by the oil production cuts. When the embargoes were lifted, after about five months, oil prices remained high and economic growth slowed.[19] So it was that 1973 came to mark the end of the golden age but analysts agree that the oil crisis was an important part of the decline rather than the cause of it; there were many economic pressures in force at that time. World inflation reached 15 per cent in 1974, by which time short-term efforts to limit price increases were dominating national economic policies.[20] Even so, there was no sudden crash. Economic growth continued more slowly between 1973 and the turn of the century but still it sustained the second-highest growth rate ever recorded. Angus Maddison refers to that period as the 'neoliberal order' that followed the golden age.[21]

Looking at the twentieth century even more broadly, we can say that before the First World War, the United Kingdom was leading the way in free trade based on the gold standard, but by the end of the Second World War the economic influence of the UK had declined and the USA had emerged as a superpower, soon rivalled and opposed by the Soviet Union through four decades of cold war. The massive nuclear arsenals on both sides virtually assured mutual destruction in the event of a real war. During that time nations could be classed economically into three groups: the industrially advanced, the developing, and the centrally planned. The last group began the transition to some sort of capitalist economy after Soviet rule collapsed at the end of the 1980s. The nations involved are still widely referred to as economies in transition.

In pre-industrial times agriculture was the main activity for all countries but when people are getting enough to eat and still have money to spend they will raise the demand for whatever manufactured goods are available. At the same time, manufacturing increases its share of employment as mechanization of farming releases farm workers for other jobs. Eventually, if incomes continue to rise, people will demand more services such as health care, education, insurance, travel and entertainment, so the service sector increases its share–which raises the need for better-educated workers. The service sector now produces more than 60 percent of global wealth and is especially important in countries that depend on tourism and offshore banking. In the Bahamas, tourism generates over 90 per cent of GDP. Agriculture is now the smallest sector globally, though it remains the main one in the Falkland Islands, Somalia and a very few other countries.[22]

During the golden age, as in any other time, the economies of nations grew at different rates and the gaps between nations widened. The GDP of thirteen countries grew at a rate of seven per cent, or even more, so it doubled twice in two decades. Most of the thirteen countries are in Asia but the list includes Malta in the Mediterranean, Botswana in Africa, and Brazil in South

America.[23]

Malta, along with Japan, went on to reach high income level, and four other economies rapidly industrialized to become the four 'Asian Tigers': Singapore, South Korea, Hong Kong and Taiwan. All of them are considered to have advanced economies today. Others in the list of thirteen lost the growth momentum and failed to catch up but they include Malaysia and Thailand, both of which would probably now be included among the 'newly industrialized countries' although that term has no agreed definition.

At first glance, it is hard to see a common factor among the thirteen countries whose economies grew so fast. A report by the Commission on Growth and Development, published by the World Bank, asserts that such rapid growth was possible only after 1950 when global trade became more open yet tightly integrated in the range of goods that could be included in trade negotiations. The fastest-growing economies all took full advantage of the opportunities; importing new ideas, technology and techniques as they monitored global demand for the products that they could supply. In addition, they all relied on the market to provide price signals and enough incentive to respond to the demands. Property rights and the mix of government interventions varied but in all cases firms and entrepreneurs could be confident that their ownership of assets was secure enough for the investments they might want to make.[24]

This sort of analysis can easily be done with hindsight but the governments of the day, if they were committed to growth, needed to have the ability to pursue it. They had to decide on acceptable levels of inflation, and avoid volatility while encouraging savings and investment in the future rather than spending. There is no universal formula for that kind of macroeconomic juggling. Global circumstances and public perceptions are constantly changing, and the same strategies can produce different outcomes in different regions and at different times.

When people are beset with feelings of economic uncertainty, for example, will they tend to save more, or spend more if they are given access to a little extra money? It is a simple question but it can trouble even the best economists when they are advising on fiscal measures involving tax rates and government spending. Manipulation of interest rates and money supply (monetary measures) are usually the responsibility of central banks or some other monetary authority.

A book described as 'one of the most influential economics books in history' was published in 1936 and, at least in the English-speaking world, the model it presented was core reading in economics courses between the late 1940s and the late 1980s.[25] The author was John Maynard Keynes (1883–1946) and the book was *The General Theory of Employment, Interest and Money*. Keynes was a product of the British Establishment but he passionately objected to the Treaty of Versailles because he could see, in 1919, that imposing punitive and unrealistic war reparations on Germany would only be counter-productive. His book on that topic was also a best-seller.[26]

One of the central arguments of *The General Theory* is that when the economy is not performing well, spending by government will have a multiplier effect provided that the recipients spend some of the money. The proportion that they spend gets passed on, accumulating through the various levels of suppliers and shoppers, and boosting economic activity. It sounds quite reasonable until we remind ourselves that it makes sense for individuals and families to spend less, not more, if they are having trouble in making ends meet. And in a domestic economy each family's spending is part of another one's income, so there could be an unintended downturn if all families began to spend a lot less. In *The General Theory*, government cut-backs would only exacerbate this, working their way through the system and causing economic activity to spiral downwards.

Keynes lost his popularity at the end of the 1970s when industrialized countries found themselves with 'stagnant' economies showing high unemployment, and continually rising prices (inflation) at the same time. It was a situation that became known as 'stagflation' and, in those days it couldn't be explained by Keynesian theory. But it was not such a puzzle for monetary theorists according to Milton Friedman (1912–2006), who was the leading proponent of monetarism at the time. Put very simply, monetarists believe the market will not need government interventions if the supply of money is properly controlled.

Friedman maintained that people do not change their spending habits just because they become a bit better-off in the short term, so government stimulus projects don't work as Keynes believed. The key to solving the inflation problem and avoiding ups and downs in employment, according to Friedman, is for central banks to grow the supply of money by a set amount each year at a rate to be decided with regard to growth in GDP. In his opposition to both state intervention and laissez faire economics, Friedman was content to call himself a 'neoliberal' but these days that is a word best avoided by anyone who is trying to be precise.[27]

Friedman won the Nobel Prize in economics in 1976, and over the years there was an impressive number of other Nobel laureates from the Chicago school of thought that he represented. He inspired the free market policies of Ronald Reagan and, at least indirectly, those of Margaret Thatcher. When he died in 2006 the *Financial Times* obituary said 'he was the last of the great economists to combine possession of a household name with the highest professional credentials.'[28] And yet monetarism did not stop governments from making fiscal interventions, nor has it swept away the influence of J. M. Keynes, whose ideas are still argued with as much enthusiasm as ever. As *The Economist* recently put it, 'Decades after its conception, Keynes' multiplier remains as relevant, and as controversial, as ever.'[29] Such is the nature of economics.

14 The rise of environmentalism

The golden age of economics propelled an army of new recruits towards environmental causes. They were troubled not by the fear of a crash but by the damage being caused by the ride. Global population in 1960 was slightly in excess of three billion and the rate of increase was rising. The conversion of ever-more marginal land for agriculture, and the physical destruction of the natural environment caused by the proliferation of buildings, roads, dams and other infrastructure was plain to see even in industrialized countries with a low population density such as Sweden.[1] Some people cared, others didn't.

In support of all the growth there was an accelerating demand for raw materials that was driving the rate of deforestation and the spread of extractive industries. By the 1960s, public concerns were being expressed about whether essential natural resources would be depleted to the point of becoming prohibitively expensive and so, in effect, used up. Sustainability would become a buzzword.

But it was pollution from pesticides that kick-started the environmental movement in the 1960s, and it was *Silent Spring*, a book by the American writer, Rachel Carson (1907–1964), that put the issue under the spotlight.[2] The crude, messy pollution of early industry was being replaced by invisible poisons that were a lot harder to detect but were no less pernicious as they accumulated in the environment. Rachel Carson was not only an acclaimed writer, she was a biologist who had worked as a science editor for the US Fish and Wildlife Service so she was well qualified to reveal the harm being done by agricultural pesticides.

Carson grew up in Pennsylvania, in the small town of Springdale where the air was often heavy with the stench rising from a factory that made glue and fertilizer from slaughtered horses. Her sister worked shifts in a coal-fired power station but Rachel escaped poverty by winning a scholarship to Pennsylvania College for Women, which is now Chatham University. *Silent Spring* was published in 1962 and by then the name of Rachel Carson was known to many with an interest in literature and/or natural history. One of her earlier books, *The Sea Around Us* (1951) had won the National Book Award and remained on the *New York Times* best seller list for 86 weeks. Clearly, the sea-loving author had found a winning formula with her skill in using an evocative literary style to convey scientific facts to the general public.

The pesticide problem

With *Silent Spring*, Carson's purpose was to explain the consequences of using pesticides on the massive scale, and in the casual manner, that had been regarded as sensible since the 1940s. She wasn't suggesting they should not be used at all. Of particularly concern were the organochlorines, the group of chemicals that includes DDT. At that time DDT was routinely sprayed from aircraft over crops and pastures to kill insect pests, as well as being used just as liberally to kill mosquitoes in those parts of the world where the insects carried the malaria parasite. As a chemical compound DDT was nothing new, having been first synthesized by a German chemist in 1874, but its insect-killing properties were not realized until the late 1930s. During WWII it was supplied in millions of aerosols for coating tents and huts, while soldiers and sailors kept themselves and their bedding constantly dusted with the stuff in order to kill lice and bed-bugs.

On the home front, families sprinkled the powder around their kitchens to get rid of flies and cockroaches. DDT was everywhere and it seemed to be indispensable. I remember it well!

I shall devote a few paragraphs to the issues raised by *Silent Spring* because they are just as relevant today and now have a wider significance. The book did not contain technical information that was new; it brought together research findings that had already been published in scientific journals. What Carson did was to use existing knowledge to reveal what, to her, was a deeply shocking state of affairs. In retrospect, it is easy to see how DDT had become so commonplace. After all, it was cheap, lethal to insects, and apparently harmless to warm-blooded animals. Millions of people had been dusting themselves and their pets with it for years without showing any ill-effects. It seemed like the perfect pesticide, although there was evidence that some insects were already evolving resistance to it.

It is a property of the organochlorines that they don't dissolve in water but are soluble in fats and oils. DDT passes through the pores in the exoskeleton of insects but is not absorbed through human skin; it most easily gets into the blood of people and other vertebrates via the digestive system. Inside an animal's body, DDT and products of its metabolism accumulate in fat and fatty parts including brain tissue. In these storage places, DDT and its metabolites can build up to concentrations hundreds of times higher than those that have been measured in the bloodstream. Ultimately, the compounds are further metabolized and released in body fluids, including mothers' milk.

So just how poisonous is DDT? For comparative purposes an acceptable, though basic, laboratory measure of acute toxicity is the median lethal dose (LD_{50}). This is the dose that kills 50 per cent of animals in a test. The LD_{50} makes statistical sense for comparing the toxicity of different substances because individuals can differ so much in their tolerance of poisons. The lethal dose for the most resistant animal in a group can be much higher than that for the most sensitive one. The reported LD_{50} tests for ingested DDT show it to be only slightly to moderately toxic to mammals. In guinea pigs, for example, the LD_{50} is 300 mg/kg (300 parts per million).[3] Many products around the home are more poisonous.

So much for acute toxicity. Chronic toxicity is vastly more complicated because it refers to any sort of harm that a substance might do to an animal when it is administered at low doses over long periods. National and international institutions, such as the US Food and Drug Administration and the OECD, maintain comprehensive guidelines for chronic toxicity testing in the laboratory. There are guidelines for revealing everything from skin irritation to chromosome aberrations in an animal's bone marrow. The problem is that years of laboratory testing of a drug or pesticide will produce findings only for a number of test species and the substance being tested. Out in the wild the substance will be one in a cocktail of thousands of chemicals, with unknown interactions between them, and the particular test species might not even exist at all.

Because DDT persists for years in soil, and for several weeks in water, and because animals store it in their tissues, it is accumulated day after day and passed up the food chains of predatory animals. In a classic study of Clear Lake in California, DDT was found in the water at a concentration of 0.02 parts per million (ppm). It had been put there to control gnats. In plankton the concentration was 5.3 ppm, and in small fish it amounted to 10 ppm. Big predatory fish contained 1,500 ppm and fish-eating birds (grebes) were being killed by a concentration slightly higher than that in the fish they were eating.[4] This sort of build-up is called biomagnification.

In *Silent Spring* Rachel Carson recounted scores of studies and observations in which biomagnification of pesticides had killed fish, mammals and birds of many kinds and in great numbers. Populations of some birds had crashed as a result of DDT contamination because their eggs were no longer hatching.[5] With regard to human health, Carson pointed out that while deaths from obvious accidents with poisons were tragic, they affected relatively few people. On the other hand, everybody was at risk from the unknown effects of a lifetime's exposure to countless combinations of chemicals that had become impossible to avoid. Unnoticeable effects on the liver, endocrine glands or reproductive cells might set off a chain reaction to produce serious symptoms many years later, or even in a future generation if human embryos were harmed by toxins from the mother. Carson had plenty of examples to make her points, though not nearly as many as she might find today. We now know, for example, that DDT and its metabolites do cross the human placenta.[6]

According to the *New York Times* the industry's response to *Silent Spring* proved more aggressive than anyone expected, with threats of legal action against both the publisher and *The New Yorker*, which had serialized the book. At the personal level, Carson was accused of being a communist sympathizer and was referred to dismissively as a spinster with an affinity for cats. But President John F. Kennedy took the warnings seriously and set up a presidential committee to investigate pesticides. In June, 1963, Carson testified before a Senate subcommittee. She wore a wig to hide her baldness and she had trouble walking because of pelvic fractures. Within the year she died from breast cancer.[7]

The use of DDT within America was banned in 1972, though it was still exported until the mid 1980s. It was not banned in the UK until 1984, by which time a lot more research had been done.

An international treaty on persistent organic pollutants was adopted in Stockholm in 2001. As a convention it came into force in 2004 with a dozen pollutants being specified for control measures.[8] Most of them were pesticides but the list also included a group of chemicals known as dioxins, which can be released during the manufacture of pesticides and from the inadequate burning of hazardous waste. A much larger group of industrial compounds, collectively known a PCBs (polychlorinated biphenyls), was also included. They have some very useful properties and until a few decades ago they were incorporated into a wide variety of products. Production has not entirely ceased, and PCBs will in any case continue to be released into the environment from the gradual breakdown of existing products, including electrical devices, plastics, paints and insulation materials.

Clamour and concern

If Rachel Carson ever thought that people would unite in sharing the views and feelings expressed in her book she would have been sadly disappointed. Long before she came to write on the subject there had been differing views of our relationship to the natural world which, in their extremes, might be described under three headings: the purely economic, the anthropocentric and the biocentric. In the first camp were those who saw the natural environment as something either useful, or if not, then as an obstacle to be overcome in the pursuit of development and economic growth. The anthropocentrics also viewed humans as being somehow set apart from the rest of nature, but not without a responsibility to look after it. In other words, humans should practice good stewardship over the rest of nature. This is broadly in accordance with the teachings of Christianity, Judaism and Islam. Buddhists and Hindus have a more biocentric tradition.[9]

Biocentrics regard humans as an integral part of the natural world, not as something independent of it. Ecologically, they are absolutely right. Indeed, the extent of our interdependence is surprising the experts. About 90 per cent of the cells in our bodies are microbial, not human, and we have long known that without the colonies of microbes with which we evolved we would have very inefficient digestive systems. But that is not the whole story. There is growing evidence that interactions between diet, gut microbes and human physiology can affect the risk of developing medical conditions as far removed from indigestion as are obesity, breast cancer and autoimmune disorders.[10] Interdependence of species, it appears, has no limits at all.

It is surely a good thing to understand the facts of life correctly, but to what extent should they dictate the moral and ethical codes by which we live? This is where science must hand over the facts to society, and for society to make of them what it will. Biocentrics consider that wild nature has its own inherent value, quite apart from any utility value it may have for humans. In other words, nature has value even if there were no humans to value it. Some philosophers agree that this is so; value has been discovered by people but humans did not create it.[11] Jainism is an ancient religion that was founded in India, and there the most devout followers try not to kill any another creature, not even an insect. In the surge of environmentalism that began in the 1960s no biocentrics, as far as I am aware, went as far as the Jains in practising their respect for all forms of life, but there are animal rights advocates with respectable credentials who argue in favour of veganism and the recognition that 'all sentient beings, human or non-human, have one right—the basic right not to be treated as the property of others'.[12]

Similarly, the term 'deep ecology' came to be used in the early 1970s for an environmental philosophy that was centred on our interconnections with the rest of nature. The movement was founded by the Norwegian philosopher Arne Naess (1912–2009). As the movement grew it attracted followers with more specific agendas such as those of eco-feminism, social ecology, social justice, bioregionalism and world peace.[13] Naess believed the principles of deep ecology were broad enough to encompass all of them but the movement was much too factional to be harmonious. Many who were swept up with the environmentalists were motivated by social and political concerns that they claimed were responsible for environmental degradation, but not all showed much interest in, or knowledge of, non-human systems.

For the American anarchist, Murray Bookchin (1925–2006), social ecology meant the doctrine that would lead to his vision of radically green, cooperative societies without any hierarchies of power. He considered deep ecology to be 'more New Age and Aquarian than the environmentalist movements it denounces under these names'.[14] Rachel Carson undoubtedly brought 'ecology' as a word and concept to the public, but it was a concept that came to mean very different things to different people. As far as popular interpretations can be lumped together at all, the word has come to refer to social behaviour and attitudes towards the natural environment, rather than the scientific study of it. I doubt there are many non-scientists for whom the word 'ecology' brings to mind thoughts of repetitive measurements, records and data analysis, but it will be readily associated with environmentalism.

I think it is also safe to say that, at least for the past few decades, environmentalists have been mostly on the left of politics and not at all happy with modern capitalism.[15] And yet environmentalists who accept that there can be natural as well as social constraints to human progress are reproached by Marxists who reject the idea of natural limits. As one Marxist writer recently put it 'a new generation of radicals who want to change the world for the better will have to challenge these left-wing greens and their naturalistic nonsense, just as surely as Marxists challenged Malthusians in the past.'[16] As I said, environmentalists are not a harmonious group, but their internal schisms are unlikely to interest those free market capitalists who regard all 'greenies' as one of a kind, the kind whose antics serve only to hold back progress and limit prosperity.

To conclude with the subject of pollution, I should emphasize that pesticides did not contaminate an entirely clean environment; the earlier industrial pollution of air and water had lessened but was still a serious matter, even in Europe and America. And the state of the great outdoors was a worry for anyone who learned enough about it, not just those with an interest in wilderness or particular species. The question whether nature had intrinsic or only instrumental value must have seemed preciously academic to many folk. Either way, something needed to be done because poisoning and pollution had become a health issue that was too big to be ignored in the twentieth century.

15 Beyond pesticides: global responses

From the late 1960s onward, new agencies were created at every level of authority from local government upwards, while special interest groups and lobby groups proliferated in their thousands. One current website declares it has links to a thousand of the most useful environment-related websites.[1] But at the global level, in the development of international law, three distinct periods have been recognized since WWII.[2]

The first period began in 1945 when the UN was set up. It had no environment or conservation unit among its specialized agencies, but both UNESCO (the UN Educational, Scientific, and Cultural Organization) and FAO (Food and Agriculture Organization) had provisions that were eventually put to use. In 1948 UNESCO sponsored a conference that produced the International Union for the Protection of Nature, a unique organization that went on to become the IUCN, based in Switzerland.

The IUCN had small beginnings and it needed to grow rapidly to keep abreast of the threats to the world's natural systems and biological diversity. But the budget was tiny. It was the shortage of funds for conservation projects that led to the creation of the World Wildlife Fund (WWF) in 1961. The Fund had the status of a charitable trust and its mission, under the now-familiar giant panda logo, was to raise money for the work of IUCN and other international conservation organizations. At that time, fund-raising was dominated by a preservationist agenda for species and habitats, and was based on popular appeal.[3]

Since then, the IUCN has expanded enormously and it now has a membership of 1,300 organizations, including some states and government agencies; individuals cannot be members. The Union has observer status at the UN, and can call on the expertise of some 16,000 experts from around the world in addition to a staff of about a thousand people working in fifty countries.[4] Through this network, the IUCN monitors the status of selected plants and animals and provides a summary of the information in the *IUCN Red List of Threatened Species*. About 80,000 species are currently listed but the *Red List* will always be a work in progress. The data are constantly being collected, revised and made freely available online.[5]

The WWF also grew swiftly and began to undertake conservation projects independently, broadening its activities into habitat conservation by helping to set up national parks and reserves. The World Wildlife Fund has retained its name in America and Canada but elsewhere it is now called the World Wide Fund for Nature. It operates everywhere as WWF under the same logo, and has become the world's biggest conservation organization with some five million supporters and about 1,300 projects ongoing in more than 100 countries. More than half its funding comes from individual donations and bequests, and as a charity it rates highly for its financial transparency and low administrative costs. As a point of interest, American donations to environmental and wildlife charities amount to only three per cent of total donations. Religious organizations receive thirty-three per cent.[6]

I should mention here the biennial reports that are produced by WWF in collaboration with a few other non-government agencies. These reports reveal trends in many dimensions of the natural environment and I shall be referring to them again. Like the *Red Lists*, they are made freely available online.[7]

Getting together and thinking big

During the 1950s and 1960s a number of treaties were adopted and there were some other institutional developments but still no organization was responsible for overall coordination of international environmental policy and law. Then, in 1968, UNESCO, with the participation of FAO and the World Health Organization (WHO), and the cooperation of IUCN, convened a conference on the conservation of the biosphere.[8] Its 20 recommendations would largely determine the themes for the 1972 UN conference on the Human Environment, commonly called the Stockholm Conference. It was this conference, attended by representatives of 114 states as well as many institutions and observers from non-government organizations, which marks the turning-point between the first and second postwar periods in the development of international environmental law.

Predictably, the conference was beset with a fundamental difference in outlook between developed and developing countries, the latter resisting any proposal that they should spend more to protect the environment as they developed their resources. Resolving this difference in the wording of the final report took days of negotiation and reworking of drafts. It had to be accepted, for example, that environmental standards that are appropriate for advanced countries 'may be inappropriate and of unwarranted social cost for the developing countries'. The Stockholm Conference did not adopt any binding obligations but it made recommendations for creating new institutions and setting up coordinating mechanisms among existing ones. It also put forward a set of guiding principles and made recommendations for future actions.[9]

At the very least, Stockholm brought the topic of global environmental protection onto the international agenda and stimulated a number of further developments. It resulted, for instance, in the creation of the UN Environment Programme (UNEP) in the late 1970s as an agency responsible only for environmental matters. UNEP now administers the Convention on International Trade in Endangered Species of Wild Fauna and Flora (CITES), which came into force in 1975. Member states (Parties to the Convention) are voluntarily bound to follow the rules in respect of international trade but national laws are still needed to control domestic trade. The Convention covers trade in thousands of species in their many forms, ranging from live specimens and body parts to items such as ornaments, fur coats and dried herbs.[10]

A few years later, in 1983, the UN General Assembly voted to create a World Commission on Environment and Development. It was chaired by Gro Harlem Brundtland, Norway's first female prime minister. The Brundtland Report, as it came to be known, was published as a paperback in 1987 with the title *Our Common Future.*[11] From the outset, the international approach to all things environmental had been solidly anthropocentric, and the Brundtland Commission consolidated that view, although the commission was independent of the UN so the wording of the report did not undergo the degree of nit-picking to which UN declarations are subjected.

The idea of sustainable development was not new but Brundtland gave the term buzzword status. It was based on an optimistic outlook for 'a new era of economic growth, one that must be based on policies that sustain and expand the environmental resource base'.[12] Since Brundtland, sustainable development has commonly been defined as development that 'meets the needs of the present without compromising the ability of future generations to meet their own needs'. But, crucially, Brundtland also pointed out that 'sustainable development can only be pursued if population size and growth are in harmony with the changing productive potential of the ecosystem.'[13] The word 'ecosystem' was used loosely but the meaning is clear enough and it is unlikely that such a straightforward statement would have appeared in a UN declaration. I shall have more to say about sustainable development in chapter 19.

After receiving the Brundtland Report The UN General Assembly called for a UN Conference on Environment and Development (UNCED). This was duly convened for 1992 in Rio de Janeiro, Brazil, and it marks the transition from the second to the third period in the development of international law. At this Earth Summit, as UNCED came to be called, 178 states were represented as well as thousands of corporations and non-government organizations. The Declaration that emerged is said to represent a series of compromises between developed and developing countries and a balance between the objectives of environmental protection and economic development. It further strengthens the anthropocentric position and emphasizes that nations are entitled to pursue their own development policies provided they don't damage the environment of other countries.[14]

The action plan resulting from the Rio Earth Summit was another non-binding blueprint for progress called *Agenda 21*, which was a reference to the 21st century. The document runs to some 350 pages and had taken more than two years of drafting prior to the conference. Many observers expressed disappointment with the outcomes of UNCED but the achievements of international organizations are notoriously limited by issues of national sovereignty so disappointment was not surprising. An agreement on actions to protect the ozone layer in 1987 was a highlight of global cooperation.

There were some tangible outcomes of UNCED. The text of a Convention on Biodiversity was adopted in 1992, having been previously prepared by UNEP. A UN Framework Convention on Climate Change was negotiated, which came into force in 1994, and a process was set in motion to draft a UN Convention to

Combat Desertification, which was also adopted in 1994.

The Convention on Biodiversity gave some impetus to increasing the number and extent of protected areas as had been called for in the Brundtland Report. Under the Convention, a protected area is a 'geographically defined area which is designated or regulated and managed to achieve specific conservation objectives' and according to those objectives the IUCN classifies protected areas in six categories.[15] Across the categories, existing areas range in size from small natural features, such as fossil beds or waterfalls, up to the Northeast Greenland National Park which is bigger than France and Italy combined. A few marine national parks are much bigger than that. We shall return to this topic in the final chapter.

From the international lawyers' point of view, despite the prevalence of non-binding agreements and lack of real progress, there was a move from setting behavioural standards towards a focus on implementation and compliance. This directed attention to such practicalities as environmental impact assessments, access to information, law-making techniques and international governance.[16]

In 2012, UNCED was revisited at another Earth Summit in Rio, attended by delegates from 190 nations including about 90 heads of state or heads of government. This summit, 20 years after the first one, is often referred to as 'Rio+20', but is officially entitled the UN Conference on Sustainable Development; the word 'environment' does not appear in the title. Deliberations largely covered the same topics as those which occupied the first Earth Summit and there was widespread dissatisfaction with the progress made on most fronts since 1992. The low priority given to protecting biodiversity was especially disappointing to some.[17] The outcome document of Rio+20 has the title *The Future We Want*, and it amounts to 49 pages in which the term 'sustainable development' appears 284 times.

Pollution

By 2017 the list of organic compounds specified in the Convention on Persistent Organic Pollutants had grown from 12 to 28, but new, and different, concerns had also grown along with the production and accumulation of plastics.

Chemically, plastics are all polymers, which is to say that they consist of long chains or networks of identical molecules (monomers). For example, ethylene is a hydrocarbon gas obtained from natural gas or petroleum, and it has simple molecules of four hydrogen atoms and two carbon atoms. The carbon atoms are connected by a double bond that can be broken with a catalyst in a reactor so that the carbon atoms connect together in long chains and the ethylene becomes polyethylene—the stuff of plastic bags and squeeze bottles.

The vast majority of plastics are made from monomers derived from fossil fuels and they can be classified in a few broad categories but the combinations of different additives and catalysts have resulted in thousands of different materials with particular properties. Natural polymers include cellulose and latex, which in the 19th century were modified to make celluloid and vulcanized rubber. Bakelite, produced early in the 20th century, was the first fully synthetic plastic but the extraordinary growth of the modern plastics industry really took off as a result of the needs and shortages created by WWII. Nowadays, indoors and on the road, we are surrounded by plastic and more likely than not we are wearing it in the form of polyester or acrylic fibres in our clothes. What we do not see are the plastic fragments that we have been eating, drinking and breathing for years.

It is estimated that 8.3 billion tonnes of plastic have been produced since 1950 and most of it has been used for packaging. As of 2015 the waste from it totalled about 6.3 billion tonnes and nearly 80 per cent of that has ended up as landfill and environmental litter.[18] Plastic litter is washed into waterways so it ultimately reaches the sea where it joins the rubbish from fishing vessels. We have long been aware of the havoc this creates among marine life, with mammals and turtles being drowned in loose fishing nets (ghost nets), and sea birds dying in their hundreds of thousands after mistaking bobbing bits of plastic for something edible. Filter-feeding shellfish are eating microscopic fragments constantly, as are small creatures that comprise the food of bigger ones and so plastic moves through the marine food webs.

There are, throughout the oceans, systems of rotating currents (gyres) that tend to accumulate floating and suspended debris. One of them, popularly known as the Great Pacific Garbage Patch, is located between Hawaii and California and is about two and a half times the size of France (1.6 million km^2). It contains debris from locations across the North Pacific and it was scientifically sampled in 2016, both from the air and by multiple vessels towing fine-mesh nets. The calculations indicated an increasing load of some 79 thousand tonnes of floating plastics within the area. At least 46 per cent of the mass of this junk was made up of fishing nets but in terms of numbers (trillions) more than 90 per cent of the fragments were classed as microplastics, being no bigger than five millimetres across.[19] Fragments smaller than half a millimetre were not quantified in the sampling, and it is believed that about 40 per cent of plastic entering the sea does not float.

Plastics are weathered away to ever-smaller pieces but they do not dissolve or evaporate and are expected to survive in the environment for centuries or possibly millennia. Nor are tiny particles produced only by the breakdown of bigger pieces; they are deliberately manufactured for particular purposes, including paints, soaps and cosmetics. There are convenient alternatives to plastic for personal care products so plastic microbeads have been banned for that purpose in many parts of the world. Some manufacturers have been pleased to show their green credentials by switching to alternatives voluntarily.

Microbeads are manufactured down to a few millionths of a metre in size but the smallest plastic particles (nanoplastics) are measured in billionths of a metre and at that size they can pass from the gut or the lungs and travel in the bloodstream and the lymphatic system to other parts of the body. The surface chemistry of particles, as well as their size, affects their capacity for moving through living tissues. Moreover, the surface properties determine which other substances from the environment might be collected and carried along with them. It is known, for instance, that PCBs and organochlorines, including DDT, will readily bind to the surface of plastic particles.[20]

Micro- and nanoplastics are released on land as well as in water, with landfill being the obvious source. The roads and open fields also produce particles from the wearing of tyres and road paint, the weathering of plastic mulch, and from sewage sludge that has been applied to the land as fertilizer. Washing machines are a substantial source of microfibres that are not all removed by wastewater treatments. Microplastics have been found in tapwater, beer, honey and household dust.

The obvious conclusion is that it has become impossible to live a normal life anywhere without ingesting plastic. The question is how much does that matter. The answers will depend on findings from future research but an editorial in the British medical journal *The Lancet* saw fit to describe it as an 'urgent problem'.[21] Plastics are here to stay and we can expect to see more piecemeal attempts at state and regional levels to reduce the growth of plastic waste, and perhaps to ban or phase out the more hazardous additives and products as they are revealed. As for a global response to the impact of micro- and nanoplastics on everyone's health, the world is only now becoming aware of the possibilities.

Biological diversity

The Convention on Biodiversity came into force in 1993 with the objectives of conserving biological diversity, using its components sustainably and sharing the benefits from genetic resources in equitable ways.[22] In 2000 the Secretariat of the Convention published a guide to explain, in simple language, what the Convention was all about, but I suspect that most people still do not have the measure of it.[23,24]

The term 'biological diversity' became contracted to 'biodiversity' in the late 1980s and it came into popular use during the following decade. Biodiversity is a collective noun for life's variety so it exists at the levels of genetics, species and entire ecosystems but it is commonly used as if it meant nothing more complicated than a simple list; a roll-call of species that can be found in this place or that. Biodiversity is infinitely more complicated than a list but it is hard to find a familiar word or phrase that would be less misleading. It is daunting to think of the awesome dynamics of even a small ecosystem if we reflect for a moment on the flow of energy through food webs, the selective pressures on genotypes, the tensions and conflicts that separate the niches of species. Darwin came up with the word 'wedging' for the tight packing of niches, and I have never felt it to be an adequate metaphor—even for that single aspect of complexity.

I have tried imagining the layer of life around the planet in terms of astronomic numbers of tensioned springs of vastly different sizes and strengths, all channelling energy from the sun and pushing or pulling against each other in such a way that the whole structure hangs together in three dimensions. If you drag a few springs out, the neighbouring ones will instantly expand into the gap. If you force one into a new position it might be squeezed out of existence or it might push its neighbours aside and become an introduced pest. Some springs are linked so that when one is gone others will surely follow. Others may serve as anchors, and whole clusters might be lost if they are removed. It is impossible to predict. Darwin's wedging, as a metaphor, might be as useful as any other, but at least the springs idea has more to offer as a mental plaything!

In real life, human activities can be in conflict with the needs of wild nature at all levels; a wetland or an entire fishery, a mountain stream or a major river, a wooded glade or a forested landscape—the list could go on as natural habitats everywhere are subjected to the proverbial death by a thousand cuts. At the species level, habitat loss tops the list of five categories of threat, the others being over-exploitation, pollution, invasive species, and climate change. Over-exploitation is not always a result of intentional harvesting. In fisheries, for example, the bycatch takes a huge toll of species that are not targeted. Similarly, invasive species may result from deliberate introductions, as with domestic cats that become feral, or from accidentally transporting species such as disease organisms.

The basic facts and figures do not necessarily convey the true measure of losses. For example, the total area of natural forest in a region may sound substantial but the extent to which it is fragmented can seriously affect its biodiversity. As a result of human activity, biodiversity is often lower near the edges of forest than in the interior, so, ecologically, very small remnants are like patches of forest edge with no interior. Globally, 70 per cent of the world's remaining forest is within one kilometre of the forest's edge, and nearly 20 per cent is within 100 metres of an edge.[25]

The argument can be made that local efforts are likely to be more effective in protecting the natural environment than are the measures taken by big international conservation organizations.[26]

But the world's biological riches are not evenly distributed around the planet; in general there is a gradient of biodiversity that diminishes from the tropics to the poles. Within this broad pattern there is a mosaic that varies from barren landscapes to regions of mega-diversity and 'biodiversity hotspots' supporting species that are found nowhere else. For conservation purposes the WWF has produced a classification scheme with 26 major habitat types further divided into 'ecoregions'. On this basis there are well over 800 ecoregions on land and hundreds more in marine and freshwater environments. In trying to protect representative examples of these it must surely help to have a coordinated global effort to determine priorities and raise funds.

Moreover, some massive impacts on the natural world result from activities such as mining and vast plantation schemes for which planning decisions are taken in the boardrooms of multinational companies. High profile, international conservation organizations are best-placed to have an influence there, despite claims that they can be bought off by big corporations who do nothing more than a little 'greenwashing' of their activities. There are successes and failures at every level but it seems clear that global, regional, national and local conservation efforts will all continue to be needed.

The first *Living Planet Report* was produced in 1998 using data that had been recorded throughout the previous thirty years or so. The report introduced the *Living Planet Index* (LPI), which has been under constant revision ever since and is now much more comprehensive. In 1998, The LPI was based on changes in the area of natural forest (corrected for plantations), and on changes in populations of sampled species of vertebrates in marine and freshwater environments around the world. The Index showed a decline of over 30 per cent between 1970 and 1995, which was interpreted as the extent of the loss of natural wealth.

But not all losses are detected in surveys of total forest area.

Partly because of fragmentation, biological richness can be lost at a much faster rate than is indicated by a decline in forest cover, and the larger edible species can have been locally exterminated by selective hunting for bushmeat—an outcome known as the 'empty forest syndrome'.[27] Unsustainable hunting, of course, is not restricted to forests. Many savannahs, grasslands and deserts in the developing world are becoming empty of harvest-sensitive wild mammals. An estimated 89,000 tonnes of meat are harvested annually in the Brazilian Amazon, where exploitation rates of large mammals are believed to be only a fifth as high as they are in the Congo basin. The availability of wild meat affects the food security of hundreds of millions of rural people around the world.[28]

The *Living Planet Index* for 1998 also contained data for the global use of six categories of resources. Four of them; grain, seafood, wood and fresh water, can be measured or estimated directly. Of the other two, carbon dioxide emissions were used as a proxy for fossil fuel consumption, and cement use was taken to be a rough indicator of the extent of building and infrastructure growth. These six components were given equal weighting and combined to give a measure of consumption pressure, nation by nation. Consumption pressure was considered to be closely related to the degradation of natural ecosystems and was found to be growing at about five per cent per year.

Possibly the six components of consumption pressure should not have been given equal weight but it was the simplest approach and at least it demonstrated how individual pressures add up with the size of population. The inhabitants of Taiwan, for example, had a personal consumption pressure of more than three times the global average while individuals in mainland China had a little less than the global average. But because of the much bigger population, the Chinese mainland showed sixteen times the consumption pressure of Taiwan. For the record, Turkey and

Ukraine were sitting exactly on the world average in terms of pressure units per person per year. Above this average there were 59 countries representing 18.4 per cent of the world's population at that time.[29]

From 2000 onward, the *Living Planet* reports included measures of the ecological footprints referred to in chapter 11, and it is now common to find them in any report, scientific, official or popular, that deals with sustainable development. It would be extremely useful to have a single number that tells us how well we are doing as a nation, or globally, in trying to live sustainably but it is not a realistic proposition. An ecological footprint number really tells us nothing about the quality of the environment that we might end up sustaining.

In working out national footprint accounts, thousands of data points are used to calculate the demand on biological resources (the consumption footprint) and the area of land needed to provide those resources and absorb the waste products from their use (the biocapacity). Obviously, the productivity of different types of land varies enormously; it may be rich cropland or poor pasture for instance, so areas for both footprint and biocapacity are expressed in abstract units called 'global hectares', which are found by averaging out the different values within the major land-use types around the world—forest, grazing land, cropland, fishing ground and built-up areas.

At the national level we find that some countries are living within their biological means. Sweden, for instance, has a footprint of about six global hectares per person and a biocapacity of about eleven global hectares. At the other extreme, Singapore has a footprint of about seven global hectares per person and has virtually no biocapacity at all. So Singapore is in 'ecological deficit' and must depend on overseas trade to meet its needs. Looked at globally, in 2012 there were 12.2 billion global hectares (gha) of biocapacity, which worked out at 1.7 gha per person.

But the world average footprint was 2.8 gha per person so there was an 'overshoot'—a deficit that could not be met with less than 1.6 planets.[30] However, nearly all that deficit was made up by one component, that is the earth's capacity to absorb our excess CO_2. Other capacities are much closer to consumption levels; after all, we can't be eating food that doesn't exist. So the footprint methodology really tells us little more than how well the earth can cope with the CO_2 load at existing levels of technology.

Indeed, as a measure of how well we are treating the natural world, the footprint accounts can actually be misleading. For example, biocapacity will increase if farming efficiency improves, and it will decrease if land productivity declines for some reason. This being so, we could reduce an ecological deficit by farming more intensively to achieve higher production. That would show up as a plus in the footprint accounts even though it may have further destroyed wild ecosystems and biodiversity. And, in principle, we could score another plus by destroying ecologically-rich old forests to make space for fast-growing eucalyptus plantations that are ecologically poor but better at soaking up CO_2.[31,32]

So while it would be helpful to be able to represent great complexity with a single number, we need to look at a few more indicators if we want a realistic overview of our impact on the planet. Why, then, has the footprint methodology become so popular? I shall try to answer that question in the next chapter.

16 Climate steals the thunder

Suspicions about human influence on climate were being voiced long before there was any scientific evidence to support them. As early as 1896 a Swedish scientist, Svante Arrhenius, calculated that increasing CO_2 in the atmosphere would bring about global warming but his work didn't arouse much interest. An English steam engineer, G. S. Callendar, was one of the few who were impressed by the idea and after analyzing existing data he published his own findings in 1938, also linking CO_2 to global warming.[1]

Until the 1950s, atmospheric gases had never been continually monitored. Even then, when the Mauna Loa Observatory in Hawaii was given the task, there was no great concern about the issue. In the report of the UN Stockholm conference of 1972, climate received only a brief mention under the general heading of pollution. Evidence for climate change was accumulating, however, and by 1979 it had became too troubling to ignore. In that year the World Meteorological Organization (WMO), in Geneva, saw fit to sponsor the First World Climate Conference.

In 1988 an Intergovernmental Panel on Climate Change (IPCC) was set up by the WMO and UNEP, and since then the IPCC has published five Assessment Reports which are claimed to be the most comprehensive scientific reports about world climate change ever produced.[2] A sixth report is expected to be ready in time for a scheduled global stocktake in 2018.

Politics, tactics and science

The steadily rising concentration of CO_2 in the atmosphere was common knowledge after two or three decades of monitoring, by which time other indicators of climate change were adding to the evidence. There are several separate indicators of global warming and that is a crucially important point. Oceans as well as the atmosphere have warmed, sea level has risen, and snow and ice have diminished; nothing is changing in the opposite direction. According to the IPCC, warming of the climate system is 'unequivocal' and it is 'extremely likely' that greenhouse gases released from human activity are the dominant cause of it. Short-term variability still exists, so over periods of a few years it is possible to choose sections of the relevant graphs that seem to buck the trend, but measurements taken at the earth's surface show each of the last three decades to have been successively warmer than any preceding decade since 1850. And it is likely that the period from 1983 to 2012 was the warmest 30-year period in the northern hemisphere for 1,400 years.[3]

A changing climate, and the extreme weather events that are associated with it, will affect all natural systems as well as agriculture, services, infrastructure and the economies of human societies everywhere. It might be thought that in the short term, and in cooler parts of the world, a gradually increasing concentration of CO_2, together with warmer growing seasons, would be good for photosynthesis and beneficial for agriculture. And there is reason to believe that some crops have shown a benefit in some northern regions, including the UK and northeast China. But climate change also affects rainfall, making seasonal weather less predictable and bringing flooding rain or droughts that can wipe out crops entirely. And if crops do benefit, so might weeds, pests and plant diseases. The consequences of continued warming beyond mid-century are predicted to be almost entirely negative for agriculture everywhere.[4,5,6]

Serious efforts to limit warming will involve major changes in the way we live. Some uncomfortable choices and economic decisions will need to be made if nations are to adapt to the changes that are probably now inevitable. So it is not surprising that reports from the IPCC have had a very mixed reception among politicians, business leaders and environmental activists who in turn feed their opinions to the popular media and so to the general public. Most people will never see the original IPCC reports, let alone the thousands of scientific papers on which they are based. This is not a satisfactory situation but neither is it easily corrected.

Measurements such as melting points, temperatures, volumes and so on, can be made with incredible accuracy, but when lots of interacting variables are involved then accurate measurements are not enough; science must then deal with probabilities and levels of uncertainty. This presents a serious problem when scientific findings must be reported in everyday language because the word 'uncertainty', like the words 'theory' and 'normal', have different meanings in science and in ordinary conversation. Physicists and statisticians might define uncertainty quite differently from each other, and people with no background in either discipline will make their own interpretations. To get around this problem, the lead authors of the fifth IPCC report were provided with guidelines so that uncertainty was at least treated consistently, and terms like 'virtually certain' (99–100% probability) and 'likely' (66–100% probability) would mean the same thing whenever they were used.

Some scientists are not convinced by the work of the IPCC and that has led to those scientists being given a disproportionate measure of publicity. Controversy appeals to journalists and producers of popular TV shows, and in the interests of 'balance' they often present the views of small minorities as if they carried comparable weight to the mainstream. The BBC has been accused of such false balance in connection with climate change.[7] Secondly, scientists who support a particular industry, ideology or political position are likely to be celebrated by those whose purposes they serve, and quite possibly they are put forward for more publicity than would otherwise be the case.

The lobby in favour of free trade and minimum government regulation is especially powerful in the USA. Think tanks such as the Heartland Institute, the Competitive Enterprise Institute and the former George C. Marshall Institute have worked hard to discredit the science behind the evidence for pesticide damage, the health hazards of tobacco smoke, acid rain, the hole in the ozone layer and now the changing climate. On the 50th anniversary of *Silent Spring* the Competitive Enterprise Institute devoted a website to Rachel Carson's achievements, declaring that she had sounded a 'false alarm'. From the viewpoint of those who oppose new regulations, that is a very useful thought to plant in people's minds because it promotes doubt. It could get people wondering whether other 'false alarms' are being sounded and whether it might be wise to ignore them all.[8]

The IPCC itself comes in for a lot of criticism, partly because it does no research or monitoring of its own but instead reviews and assesses the scientific and technical publications of thousands of researchers from around the world. For this reason, say critics, the IPCC reports are more political than scientific because, in the usual way of UN reports, the wording gets thrashed out in all-night plenary sessions until the member states are satisfied with it.[9] There may well be some truth in that, but much political influence will be moderating, rather than alarmist with regard to climate change. After all, big oil-producing nations are also members of the IPCC.[10] Political influence can work both ways.

So what are people to think about climate change? Not wishing to add to the clamour myself, I will say only that when nearly all the world's specialists on the subject say that the climate is warming, and that greenhouse gases released from human activity are a major cause of it, then that is the way to bet. As to the consequences of climate change over the coming decades, in general terms, global warming will continue to make the oceans rise as the water expands and more ice melts into it. Extreme weather events are likely to become more frequent and less predictable, and seawater will become more acidic to the extent that it continues to absorb more CO_2. Beyond these effects that are linked directly with warming, the interactive consequences of climate change are so difficult to predict in any detail that I wouldn't bet on any of them. Even major outcomes are unpredicitable; how will the ocean conveyor belt (chapter 3) respond to the changing climate? Nobody knows. How much methane and carbon dioxide will be released from melted permafrost this century? Again, nobody can be sure, although recent findings do indicate that the release is likely to be 'gradual and sustained rather than abrupt and massive'.[11] It is a question of critical importance because greenhouse gases released from such a vast area of land could cause warming that would accelerate further thawing and so on in a positive feedback loop. It would be a tipping point and would inexorably run its course.

Global warming as a symptom

Global warming is but one of many symptoms of the same condition, but what condition is it? Those who embrace Lovelock's Gaia hypothesis may be willing to talk of a malady or sickness of some sort but this implies that the earth is suffering and needs to be nursed back to health, and that is altogether too fanciful. For those who like a medical metaphor, we could say more literally that the earth has got a growth. It is a big growth and a number of the earth's system settings are being automatically changed by it. They are simply changes for the planet, neither good nor bad, but it is hard to see how they could be good for people or any other form of life that has reached its present state during the relative stability of the Holocene.

Neither is it immediately obvious why global warming is so consistently discussed in the media as if it were the problem itself rather than one symptom (albeit an important one) of the growth that causes it. This is not just a quibble about vagueness or lax journalism; global warming has acquired the status of an overriding problem and as such it carries the implication that all would be well if it could somehow be reversed. It suggests that if we could solve the big problem of climate there would be no more failing water supplies, disappearing forests and wetlands, soil salinity, poor air quality, fisheries decline, disruption of nutrient cycles, extinctions and so on. Global warming is likely to amplify or complicate these and many other environmental challenges that we face, but with or without global warming, on present trends those challenges will still be there—and growing.

In 2009, Richard Black, who at that time was environment correspondent for the BBC, asked the questions: 'has climate change hijacked the wider environmental agenda? If so, why? And does it matter?' Among the replies, a former leader of a climate-change research centre suggested that politicians tend to be enticed by long-term issues of a global nature because voters won't expect a quick fix from national governments. Decisive action can be repeatedly postponed and failures can be blamed on events beyond the government's control. Britain's government very publicly made a high priority of tackling global warming. As the head of a large British environment group remarked, 'If we want to talk about climate change, we can get a meeting with the prime minister. If we want to talk about biodiversity, we can't even get a meeting with the environment secretary.'[12]

Very few political leaders show much understanding of ecology but they soon become familiar with the language of economics and have no difficulty seeing the economic consequences of reducing carbon emissions. Whether they are oil-producing, coal-producing, keen to develop renewable energy industries, or just sensitive to energy costs and rising sea levels, political leaders need no persuasion to take an interest in climate change. And once there is widespread interest at the top, it makes sense for environmental groups to play up the importance of climate change for their own cause. All causes need money.

The mass media, inevitably, pass on the apparent hierarchy of concerns if only by the number of articles and documentaries they devote to different issues. Richard Black once added up the number of articles written for the BBC News website over a nine-month period. The scores were four for deforestation, four for desertification and 17 for biodiversity. On climate change he stopped counting when he reached 1,000.[13] We have become accustomed to hearing references to 'the environment' when all the speakers have in mind is greenhouse gas.

In view of this, the popularity of ecological footprint accounting is not such a puzzle. With environmental thinking so thoroughly centred on CO_2, few will question it when a single number, derived from the area of forest needed to absorb our output of the gas, is passed off as a measure of environmental impact.

Returning finally to global responses, The United Nations Framework Convention on Climate Change (UNFCCC) came into force in 1994 and within a year governments had started to work on a protocol for reducing greenhouse gas emissions. The text of it was adopted in Kyoto, Japan, in 1997 and it entered into force in 2005. Nations signing up to the Kyoto Protocol made different commitments to reducing their gas emissions, and various mechanisms were approved for making the targets easier to reach. If a nation pays for emission reduction elsewhere, for example, it will earn carbon credits that count towards meeting the national target at home. China did not participate in the Protocol, and the USA subsequently withdrew support for it.

Looking back from 2018, it is arguable whether the Kyoto Protocol had much success because the figures show very mixed results and, overall, global emissions have risen. The 38 developed countries involved did meet their targets, though ten of them depended on carbon credits. Their emissions were also curtailed by the global financial crisis of 2008 so it is unclear how much can be attributed to the Protocol.[14] Kyoto, however, did help to pave the way to Paris.

The UNFCCC negotiated the Paris Agreement during the 21st Conference of Parties in 2015, and it went into effect in November 2016, by which time 194 nations had signed up to it, including the USA and China. Briefly, the aims of the agreement are to hold the increase in global average temperature to well below 2° C above pre-industrial levels, while making an effort to hold the increase below 1.5° C. Additionally, the aim is to become more resilient to climate change and better able to adapt to its adverse impacts without threatening food production. Critics complain that the agreement is based on nothing more binding than promises and good intentions, but they must accept that it was the best deal that could be achieved at the time.

And only time will tell how effective the agreement will be, even if the good intentions are carried out. Global surface temperature is a good indicator to use in setting a target but scientists are not claiming that a 2° C rise in average temperature will be safe. According to some specialists we have probably already passed the safe limit. Computer modelling suggests that between the years 2000–2010 global warming could have been offset by volcanic eruptions by as much as 25 per cent.[15] The limit of 2° C was a decision made by the UNFCCC that was informed by science but had to be a compromise between what was wanted and what might be achievable.[16]

Big oil companies keep a very sensitive finger on the pulse of the energy markets, and BP has been openly publishing its much-cited *Statistical Review of World Energy* every year since 1956. It provides a solid database on which to base expectations for the future, though of course not even BP can be sure what tomorrow will bring. Nevertheless, a good database is sound support for educated guesses, and since 2011 BP has also made public its annual *Energy Outlook.* For what it is worth, the outlook from 2018 shows world GDP more than doubling by 2040 but energy demand increasing by a third over the same period as a consequence of accelerating gains in energy efficiency—that is the energy used per unit of GDP. Despite improved efficiencies and a slowing in the growth of carbon emissions, the use of 33 per cent more energy is still predicted to cause about a ten per cent increase in CO_2 emissions by 2040 under existing targets.

Nearly all the growth will take place in emerging economies. The number of private and commercial vehicles illustrates the situation very clearly. In mature economies vehicle ownership is close to saturation so most vehicles sold will replace existing ones, but the total number of vehicles in the world is expected to nearly double to reach around two billion in 2040. Similarly, much of the growth in electricity generation will take place in regions where large proportions of the population are without electricity at present.[17]

Efforts to limit emissions must be looked at globally; every tonne of CO_2 has the same effect in the atmosphere regardless of where or how it was emitted. But beyond certain thresholds, a change in global temperature can have quite disproportionate effects in different regions, and sorting out the complexities of it all has necessarily become the work of specialists who develop global climate models on computers. A global 2° C rise, for instance, might mean a warming of 5° C in some areas, but also depending on locality, it could lead to local rainfall being either reduced or increased by 20 to 30 per cent. The uncertainty increases with the degree of climate change. The continuing challenge for the nations of the world is to decide on global targets when it can be seen that risks are high in some places and low in others.[18] Science can be expected only to provide a more solid foundation on which to base such difficult decisions.

17 Population

From an ecological point of view, as I mentioned earlier, populations, are living systems at one of three main levels of organization, the others being the individual and the community. A stable population is one that is neither increasing nor decreasing but still it must be in a state of constant change to some extent. Babies are being born, there will be deaths, and there may be arrivals and departures of migrants. Any population will stay the same size for as long as numbers of births, deaths and migrations cancel each other out. Those three demographic processes, fertility, mortality and migration, are the only ones that can change a population's size, distribution and structure. Population structure refers to the proportions of males, females and different age groups.

In addition to these ecological basics, human populations are affected by many interacting and uniquely human variables such as occupation, health, wealth and education, that are influenced by the type of society and cultural behaviour of individuals. The raw data are collected during censuses and other surveys, and it is the work of demographers to make statistical sense of them all. In this chapter I shall be referring only to human populations.

Population size and structure

Population structure is most often shown in the form of a bar graph in which the bars, as age bands, lie horizontally and are stacked up with the youngest at the base and the oldest group at the top. The length of each bar indicates the number (or percentage) of individuals in that age band, so for a strongly growing population the graph has the shape of a pyramid, with infants at the base forming the longest bar and the oldest folk gathered in the shortest bar at the tip.

The proportion of males to females in each age band is shown by collecting males on one side (customarily the left) and females on the other so that each bar is divided down the middle of the pyramid to make a double bar graph. In humans the sex ratio at birth is slightly biased towards males and is often reported to average 105 males for 100 females. This is now known to be a result of slightly higher female mortality during gestation, not a bias at conception.[1] As children become adults there is usually higher mortality among males so that the female side of the bar chart becomes relatively wider. It is often conspicuously so in old age because women tend to live longer than men, but it is easy to find exceptions to this general rule because local or regional issues, such as war or deprivation, can affect mortality of the sexes differently.

Double bar graphs of population structure are probably most often called population pyramids although the pyramid shape is formed only by plotting graphs from strongly-growing populations—when each bar will be longer than the one above it. Shrinking populations and those with zero growth will produce a different shape. Some people prefer to call the graphs 'population profiles'.

Regardless of what we call it, a population pyramid is a snapshot of a population structure at a particular time but it also reveals something of the history of the population over several decades, and it indicates how that population is most likely to change in size and composition in the near future. We can visualize the age bars moving upwards through the graph so that in 80 years' time the bottom bar will have arrived at the top and become much shorter. We can see the proportion of individuals in the pyramid who are now of reproductive age and who will produce the next bar at the bottom, although we cannot be sure how much reproducing they will do because the future is always uncertain.

However, when a massive proportion of the pyramid is made up of young people there is obviously a huge potential for future reproduction. Such a population has its own momentum and can keep on increasing for as much as three generations even if fertility is at or below replacement level.[2] Replacement level fertility is the level of fertility at which a population will exactly replace itself from one generation to the next. It will eventually lead to zero population growth (if mortality rates remain constant and there is no effect from migration), but the momentum first has to work itself out through the age structure. In developed countries replacement level fertility averages about 2.1 children per woman. It is likely to be higher in countries with high infant and child mortality rates.

In the majority of published graphs the bars represent five-year age categories (0–4, 5–9, etc.) but more detailed information is revealed by graphs in which each bar represents a single year. On the pyramid for Japan in 2006, for example, there is a noticeably small percentage of 40-year-olds. That age corresponds to the birth year of 1966, which was the year of the Fire Horse in east Asian cultures. It comes around every 60 years and is not a good time to be having babies because it is believed to be an unlucky birth year for girls. On a graph with five-year age categories the effect of the Fire Horse would be masked to some extent by births in adjacent years.[3]

World population change

The pattern of high fertility and high mortality that characterized populations of pre-industrial times began to change after mortality rates fell in response to hygiene, disease control and better health care. We have seen that populations increased enormously before fertility rates also began to fall and bring down the rate of population growth. The full sequence of events, from the pre-industrial pattern through to the end of high population growth, is known as the demographic transition and it typically takes more than a century to complete.

The demographic transition has long been recognized and debated but the model still has little explanatory power. The demographer Dudley Kirk, with fifty years of hindsight, concluded that 'it has not been resolved whether the demographic transition is a theory, a generalization ... or a mere descriptive term'.[4] The problems are that the stages of the transition have different durations in different countries, and there is no consensus as to what, exactly, triggers the stages anywhere. In the countries that were first industrialized the demographic transition has run its course but African, Asian and Latin American countries began their transitions later, and a few of them are still at a very early stage. However, the timing of the onset is not the only difference between the demographic transitions of developed and developing nations; some observers have described them as two separate models.

In particular, the better health technology, higher incomes and improved education that contribute to the lowering of mortality rates were not quickly achieved during the late nineteenth and early twentieth centuries. But the medical technologies and public health measures that became widespread after WWII were transferred relatively swiftly to developing countries, which helps to explain the more rapid decline of mortality in less developed parts of the world.[5,6] And because there was no corresponding decline in fertility rates, the populations of developing countries grew two or three times faster than European nations had ever done. During the 1960s many developing countries had rates of increase as high as three per cent a year—which doubles a population in 23 years. A few countries, notably Uganda and Niger, are still in this phase. In Niger, women have an average of seven children and more than a quarter of the girls are married before the age of fifteen. As one reporter described it, 'there are kids everywhere. Even the children have children.'[7]

The reasons why fertility does eventually begin to decline are matters of debate between sociologists and economists, and the exact combination of causes may differ within, as well as between, developed and developing nations. Most explanations put forward to explain declining fertility in the developed world are economic arguments centred on the 'price' of children as measured both in the direct costs of raising them, and in the opportunity cost to the mother when having children means going without paid employment. But while it is easy to see associations between the price of children and fertility decline, attempts to show the extent to which one causes the other have had mixed results.[8]

In order to limit family size, couples need to have the capability as well as the intention so the effectiveness of birth control practices and the availability of contraceptives could have been highly significant at different times and within different cultures. Some economic theorists have completely disregarded this point as though choosing a number of children has only ever been an economic decision equally open to every couple.[9]

It can be assumed that the earliest influences on fertility would differ from those that are now important in advanced economies, and some of those earlier influences such as the need for supportive children in one's old age, may still be strong in developing nations. Again, clear evidence of cause and effect is hard to find. There is compelling evidence that the education of girls and women leads to their having fewer children as well as improving the quality of life for mothers, but the provision of schooling is normally part of wider development plans and is not easily implemented separately. Nor is it a prerequisite for fertility reduction; there are numerous examples to show that family planning programmes in developing countries can be highly effective in reducing fertility and improving the health and welfare of mothers and children among the rural poor.[10,11,12,13]

Whatever may or may not have influenced human fertility since the end of WWII, the outcome in numbers is plain to see. In 1950 the world had about two and a half billion people compared with more than seven and a half billion today. The people of Asia comprised 55.4 per cent of the global population in 1950, while Europe had 21.6 per cent, the Americas 13.4 per cent, and Africa 9.1 per cent.[14] In 2017, sixty per cent of people lived in Asia, 15 per cent in the Americas, and 10 per cent in Europe. Africa has 17 per cent of the present total, which means there are nearly six times more people in Africa than there were in 1950.[15]

In contrast, fertility in all countries of the European Union is now below the replacement level of around 2.1 children per woman, and in most cases it has been so for several decades, but population momentum has maintained a slow increase in the Union's population as a whole. This will continue for a few more decades in some countries, though a slow decline has begun in others, especially, though not exclusively, in Eastern Europe. In 2016, for example, both Italy and Portugal recorded eleven deaths and eight births per thousand people.[16]

Future projections

Because of the huge momentum that is already built in to the populations of developing countries a substantial increase in global numbers is inevitable unless there are completely unforeseen events. In the most recent projections by the United Nations there is a 95 per cent degree of confidence that in 2050 there will be a population of between 9.4 and 10.2 billion, with the medium projection showing 9.8 billion. By that time there will be eleven times more people in Africa than were there a century earlier, and they will make up 26 per cent of the total. Fifty four per cent of people will be living in Asia, and 12.4 per cent in the Americas. Europe's population will amount to only about seven per cent of the 9.8 billion.[17]

With less certainty we can try to look further into the future. It is possible, though not probable, that the global population will stabilize or start to fall before 2100, by which time it will have reached 11.2 billion in the medium variant. Should fertility decline more slowly than the projected rate, then the numbers will be higher. For example, if all countries had a fertility rate that was just half a child higher than in the medium variant it would add five billion people to the total in 2100.[18] All projections are based on assumptions so only time will tell, but the UN population projections have a good track record over several decades. The population size for the year 2000, as projected in 1958, turned out to be wrong by only 3.6 per cent.[19]

In 2007 the urban population of the world exceeded the rural population for the first time in history. By 2014, 54 per cent of people were living in urban areas, including smaller towns of less than half a million inhabitants. According to UN projections, two thirds of the global population will be urban by 2050. One in eight people presently lives in one of the 28 mega-cities of 10 million or more inhabitants, and it is projected that there will be 41 such 'urban agglomerations' by 2030. Africa and Asia, the least urbanized regions at present, will urbanize most rapidly, but at 56 and 64 per cent respectively, they will still be less urbanized than other regions by mid-century.[20]

So far, rapid urbanization, especially in sub-Saharan Africa, has been characterized by extensive urban sprawl and serious environmental deterioration extending far beyond the urban fringes. Areas least valued for formal development, such as floodplains and unstable slopes, soon become filled with makeshift homes. It is reported, for instance, that 70 per cent of people in Lagos, Nigeria, already live in slums that are vulnerable to flooding. And on present trends, most people in Kampala, capital of Uganda, will live in flood-prone areas by 2030 and be suffering the consequences of poor sanitation.[21]

Wood and charcoal account for 80 per cent of household fuel consumption in Africa, and with urbanization the trend is to move away from wood in favour of charcoal that is often trucked into towns over distances of hundreds of kilometres.[22] Charcoal has a higher energy density than the wood from which it is made but by volume a bag of charcoal will represent something closer to two bags of wood. It is a situation reminiscent of the growth of towns in pre-industrial times (chapter 8).

Population, economy and society

Micro-economists well understand that we all have preferences and if we can get more of what we like without losing anything we will generally feel better-off. In short, more is usually better. I have long suspected that a similar principle is operating when we accept without thought that there is strength in numbers. Before there were economic theories, consumer preferences and market manipulators there were raiders and defenders. Who, in history, has ever rejoiced in having smaller numbers on the home side? Long-established ways of thinking, or not thinking, can be very slow to change.

The ageing of populations

In economics and the social sciences it is customary to regard the years between the ages of 15 and 64 as the productive time of life. On that basis, children of 14 and younger, and adults of 65 and older, are classed as dependants. The number of young and old dependants per 100 persons in their productive years defines the 'youth dependency ratio' and the 'old-age dependency ratio' respectively. The 'total dependency ratio' is found simply by adding the number of young and old dependants. These ratios can be nothing more than an approximation to reality and it can be argued that they are now outdated and misleading, but I will leave that aside for the moment.[23]

Periods of high fertility increase the level of youth dependency in a population but if the children can be usefully employed as they grow up they will become an economic dividend by expanding the labour force. On the other hand, if there are no job opportunities for them, child dependants will be destined to swell the ranks of unemployed youth. According to the International Labour Organization (ILO) the number in the category of unemployed youth, defined as 15–24 years of age, was 71 million in 2016, with a further 156 million working in 'extreme or moderate poverty'.[24] That situation is not likely to improve soon; the immediate outlook is for global unemployment to rise a little.[25] Sweeping statements to the effect that high fertility yields an economic dividend evidently need to be treated cautiously.

In developed nations there was a fertility increase, popularly known as the 'baby boom', that began after WWII and in the USA lasted for most of two decades. The duration and magnitude varied among other western nations.[26] Japan's baby boom lasted only from 1947 to 1949, though it was followed by a smaller 'echo boom' from 1971 to 1974. Regardless of whether a baby boom yields an economic dividend, the youngsters will age and the bars that represent them on the population pyramid will move upwards and eventually arrive at the post-productive years. The surviving baby boomers will then be classed as old-age dependants, and this brings us to an issue that is constantly in the news of advanced economies at present—the problem of population ageing.

Ageing and life expectancy

High population growth will always be followed by an increasing old-age dependency ratio as the demographic transition runs its course and the last of the fertility bulge approaches the upper part of the population pyramid. It is simply a matter of time; in some nations, notably Japan, the process is well advanced while the populations of countries in Sub-Saharan Africa and parts of the Middle East have not yet begun to age.

In a stable population, without migrants moving in or out, the population pyramid would be shaped like a column, with sides roughly parallel from the base to the advanced years. Near the top, where older people reach their age limits and die, the column would narrow to a point. There would be neither booms nor busts of babies so the old-age dependency ratio would remain constant, as would the total dependency ratio. Deaths, in this hypothetical population, would be balanced by births, no more and no less. In real life there may be periods of such stability but no populations remain absolutely stable for long periods because environmental influences on population growth are always changing. With human populations there are also the uniquely human influences.[27]

Although falling fertility has been the main cause of population ageing, the fact that people are also living longer does add to that demographic change and it becomes relatively more important after populations have had long periods of low fertility.[28] At present, life expectancy at birth for females in the more advanced economies ranges from 77 years in Russia to 87 in Japan but among less developed countries life expectancies are lower and there is much wider variation. However, the gap between life expectancies in different regions of the world has been closing in recent decades and is expected to diminish further during this century. By 2045–2050, according to UN projections, people in the more developed regions will have an average life expectancy at birth of 82 years compared with 75 years in less developed regions—a gap of seven years.[29]

Within biological limits, life expectancy automatically increases with age because fewer of the main hazards lie ahead. In other words, the longer we survive, the longer we can expect to live. For example, a boy born in the USA in 2010 had a life expectancy of 76.2 years, but a man aged 65 on that date could expect to live for another 17.7 years and so reach the age of 82.7.[30]

The ageing of human populations has no historical precedent so it has never had to be managed before. Welfare and other social systems were developed while populations were growing strongly and they were based on the dependency ratios in place at the time. Yet the ratios were bound to change and have already changed significantly. In 1950 in the more developed regions there were 82 people of productive age for every ten people over the age of 64. In the year 2000 the number had fallen from 82 to 47, and by 2050 there will only 22 working-age supporters for every ten old-age dependants.[31]

The obvious weakness of these data lies in the arbitrary definition of productive years; individuals vary enormously in that regard, as do the physical and mental demands of different occupations. We will return to that shortly. In general, population ageing presents economists and politicians with a serious challenge that is often reported to the public in sensational terms; serious newspapers in Britain having seen fit to refer to it as a 'demographic time bomb' and a 'population disaster'. Italy was declared to be 'a dying country' by the nation's health minister; and Vitor Constâncio, vice president of the European Central Bank, described the falling birth rate as 'collective demographic suicide'.[32]

In 2013 the Pew Research Centre conducted a survey of attitudes towards ageing in 21 countries around the world and found that the level of concern among the general public roughly tracked the extent of ageing. East Asians were by far the most worried, followed by Europeans. The American population is ageing more slowly than those of Europe, and only one in four Americans believed ageing to be 'a major problem'. Notably, America was one of very few countries where a wide section of the public believed individuals should be responsible for their own well-being in old age. In 13 of the 21 countries surveyed most respondents thought government should bear that responsibility, though many people were doubtful about the standard of living they were likely to experience.[33]

Responses to ageing

There are both social and economic consequences of population ageing. A concern of national governments is that the pool of workers who provide tax revenue will be shrinking while the costs of health-care and pensions will be growing. Economies must also respond to the changing demand for goods and services as populations age towards the 'silver economy', and economists fear its effect on the growth of GDP.[34]

Accordingly, governments of developed nations have put a variety of expensive measures in place in an attempt to increase fertility. Baby bonuses, paid parental leave and child-care subsidies for the first two babies can all be regarded as part of a welfare system but when they are provided (and more generously in some countries) for a third and subsequent child it can be with the intention, expressed or otherwise, of boosting population growth.

Some attempts to promote procreation seem to be born of desperation. In one region of Russia, on the twelfth of September, couples are given time off work so that they can go home and try for a conception. Women who give birth exactly nine months later are rewarded with a prize on Russia's national day—June twelfth. Similar forthright schemes have been reported from a number of countries.[35] Italy's campaign for more babies included the release of a series of online images urging women to get pregnant, though the role of men was not overlooked. One image showed a man's hand holding a limp and smouldering cigarette with the slogan 'don't let your sperm go up in smoke.'[36] Italy's 'Fertility Day' images were withdrawn after they proved to be more offensive than persuasive but other strategies might work in the way that governments hope.

France has a range of policies to benefit families and especially families of three or more children. The average number of children born to women in France is close to two, which is a higher fertility rate than exists in other European countries, but it is difficult if not impossible to assess how much this is influenced by any particular policy. Nevertheless, research by Olivier Thévenon suggests that, collectively, the policies for providing parents with services and money for raising children do have a positive impact on fertility.[37] It could be, however, that different socio-economic groups respond differently to specific policies. In a survey carried out among American college students the respondents associating themselves with the poorest 'economic class' were more likely to choose daycare or money as the most valued incentive.[38]

Among OECD countries there is a lot of variation in personal opinions about the ideal family size. In Austria the preferred number of children, on average, is below replacement level but the average in other countries works out between two and three children. The people of Ireland have the highest average ideal number; women opting for 2.8 children, and men 2.6.[39]

The other major strategy used by governments to counter low birth rates is to boost the number of working-age immigrants. This is a short-term strategy with immediate economic appeal. It will at once increase the size of the work force, and for a time it may increase the fertility rate if the immigrants are accustomed to bigger families in their countries of origin.

But migrants' birth rates generally conform to those of the host country after a few decades, and migrants also grow old so the government, or more to the point, a later government, will simply end up with a bigger ageing population when growth ceases. To make matters worse, as a result of working migrants being already considerably older than newborn babies, 'a steady stream of migrants almost always makes a population younger in the short-term but older in the long-term, as compared to the age structure in the absence of migrants.'[40] Exceptions to this rule are mathematically possible. A receiving population could already be so old that the average age of migrants, during the time they lived as part of it, would age it no further.

The unavoidable fact is that populations cannot be kept growing forever, either by boosting fertility or promoting immigration, so societies must adapt to the new normal that will be left after the demographic bulge has passed through. But postponing the inevitable does allow a population to get bigger for a time so it is conspicuously favoured by those sectors of society that profit most from increased population density—even when it lowers the quality of life for most people.

What, then, can be done to maintain a high standard of living in an ageing population, given that panic measures are not helpful. The question can be tackled on several fronts at different levels but some answers can be gathered under a few headings. For a start, the figures that we see reported can make the problem seem bigger than it really is. The 'burden' of old dependants is offset to some extent by there being fewer young dependants. There is also a gain from the inheritance of capital stock insofar as a falling birth rate results in fewer children having to share the inheritance from the previous generation. But, most importantly, the challenge is grossly overstated by those who assume that we must have a fixed retirement age and that older people must inevitably cost more in health care.

In reality, people enter and leave the work force at different ages and on different terms—part time or full time and sometimes retiring in advance of pensionable age. This flexibility can continue provided that pensionable age increases along with longevity. If pensionable age were advanced to equal the increase in longevity the age dependency ratio would be reduced. Maintaining the same proportion of working years to retirement years keeps the ratio unchanged.[41]

But there is a widespread perception that older workers are generally less able than younger ones, regardless of their knowledge and skills. Both private and public-sector employers may still have pension schemes with incentives to take early retirement rather than having retraining schemes designed to keep older workers in positions that are not physically too demanding. On the other hand, if Finnish experience is any indication, more flexible pension schemes do not necessarily have the desired objectives either.[42]

With regard to health care, the available evidence indicates that the costs of aged care are not increased as a result of people living longer; rather the same costs arise at progressively older ages, ending up with a very large proportion of the total cost being incurred during the last year of life.[43,44] A study in Australia also showed that, in that final year, hospital costs were actually lower for older people. The average cost per person for people dying between the ages of 65 and 74 was more than twice as high as that for a person who died at 95 or over.[45] When findings such as these are taken into account it becomes evident that measurements of dependency entirely based on fixed age groups are not at all appropriate for planning purposes.

A different measure could be based on the expected years remaining at a given time rather than on the years lived. In 1950, for example, a 65-year-old woman in the USA or Sweden could expect to live for another 15 years or so, but in 2014 the equivalent expectation was 21.6 years in Sweden and 20.5 years in America.[46,47] Two leading researchers in this area, recognizing that there are backward-looking and forward-looking views of ageing, introduced the term 'retrospective age' for time lived, and 'prospective age' for time remaining. They point out that many of life's decisions, such as whether to learn a new skill, or whether it would be worth having a hip replacement, depend on reasonable expectations of future activity rather than on the chronological age that dominates present thinking about population ageing.[48]

There is no doubt that population ageing does present challenges but they are largely a consequence of social systems having failed to adjust to ageing in a timely way so there is enormous scope for policy improvement. Beyond that, the key to a longer working life (and longer retirement) is healthy ageing, and that is something that we can at least try to improve for ourselves. At the institutional level, Europe's 'Covenant on Demographic Change' has been set up specifically to promote healthy ageing and will monitor progress with a newly-created 'Active Ageing Index'.[49,50]

All in all, the problem of population ageing in wealthier countries has been overstated. Perhaps more importantly, it may have diverted attention from the far more serious problem of continued high population growth in poor countries, many of which have scarcely any institutional capacity to cope with yet another doubling of numbers.

18 Numbers and the green revolution

The relationship between human beings and the natural environment has become indescribably complicated, although there are clear examples of incompatibility as in the case of people and lions in Africa. Lions are not too difficult to kill with combinations of poisoning, trapping and shooting, and they do not survive where human population density has reached about 25 people per square kilometre. Where that density has been reached, lions and people with their associated livestock do not coexist.[1]

Direct conflict is obvious and easy to understand but much of our impact on the natural environment is neither direct nor intentional. If the broad picture can be framed at all it is by the well-known equation in which the aggregated impact (I) results from the number of people (P) multiplied by their consumption of resources (A for affluence) multiplied by their level of technology (T), or in short, I = PAT.[2]

Think of volume, which is found by multiplying length times width times height. It makes no difference whether the three dimensions are all the same or are very different; a cube can have the same volume as a shoebox. Similarly, within limits, comparable damage to the environment can result from many people with primitive tools or a few wealthy people with machinery; a few rich and wasteful consumers or a lot of hungry people trying to eke out a living. However, volume is volume no matter how it is shaped whereas environmental impact includes damage of all kinds.

An average person in an industrial country puts much more pressure on natural resources than does the average individual in a poor agrarian economy. In consumer societies it is easy to see the impact of wealthy people with modern technology. There is massive damage both from the consumption of natural resources and the inexcusable amount of waste that results. If wealthy societies lived more moderately, some would say more sensibly, it would certainly make a big difference. Good examples are urgently needed to counter the immense growth in consumerism that is now under way among expanding populations, especially in Asia.

In China, for example, where we have already noted the growing number of millionaires, online shoppers spent a record US$25 billion on Singles Day in 2017.[3] This dwarfed American spending over the Thanksgiving weekend. As Deloitte describes it, young adults who make up a quarter of the population in China are now aggressively using online retailing as well as increasing their demand for luxury items such as cars. Indian households are no longer allowing traditional restraint to stand in the way of 'rampant consumerism', while in Indonesia the proportion of the population classified as either high-income or upper middle-income has risen from 15 percent to 60 percent over the ten-year period to 2015—an indication of consumer potential in that country.[4] The people of these nations understandably feel they have every right to follow in the path of the affluent West.

In the I = PAT equation western nations have a high score on all the multipliers involved. In contrast, Ethiopia stands out as an example of a nation with high population but low levels of affluence and technology. Twice the size of France, more than half as big again as Texas, Ethiopia is a land of exceptional diversity. Rugged mountains are riven by the Great Rift Valley with its lakes each having their own distinct characteristics. To the east are deserts extending to the Red Sea and across Somalia to the Indian Ocean. In the south-west the natural vegetation is dense forest with seasonal wetlands bordering the Sudan. It is a land that ranges in elevation from salt flats lying more than 100 metres below sea level to mountain peaks at 4,500 metres.

For thousands of years the highlands have supported subsistence farmers while in the dry lowlands nomadic pastoralists have herded their stock, all of them following ancient cultures and using traditional technology. Travellers in the late nineteenth century described a richness of vegetation and conspicuous wildlife that can now be found only as remnants. In 1900 the human population of Ethiopia is estimated to have been about twelve million. The number had grown to about eighteen million by 1950. At the Stockholm (1972) Conference on the Human Environment, Ethiopia was one of a dozen nations that voted to delete a recommendation on the need for family planning.[5] But by 1993, when Ethiopia's population had reached 53 million, the new government introduced its own population policy.

The policy document runs to some forty pages and points out that the estimated amount of forest cover had fallen from about 40 per cent in 1900 to about 3 per cent. In a country with no coal, and where wood and other biomass is the main source of energy for cooking and domestic heating, the concern is easy to understand. So, too, is the alarm caused by estimates of more than a billion cubic metres of soil being lost from highlands by erosion each year.[6,7] Irrigation in the lowlands is still limited in scale but has brought salt to the surface and revealed the risks of salinization.[8]

The population policy has sound and clearly-stated objectives with high priority given to raising the economic and social status of women and expanding family planning through community-based outreach services. Specifically, there was a target of reducing the total fertility rate of over seven children per woman to approximately four by the year 2015. The policy evidently achieved its target but it was too late to prevent the build-up of enormous population momentum. Ethiopia's population has now passed 100 million and is projected to reach 191 million by 2050.[9] Food supplies have been increased mainly by expanding the area under cultivation. A study of land-cover change between 1977 and 2000 showed that 'areas covered by bare rock, soil and devoid of any vegetation had increased by 86.5 per cent, swamp area has shrunk by 51 per cent ... and shrub land has decreased by 83 per cent'.[10] It is reported that nearly one in five households now burn dung cakes as fuel for cooking.[11]

This devastating environmental impact could have been mitigated or avoided by better governance, different institutional arrangements and a corresponding fall in fertility in the past. But as things are, Ethiopia provides a clear example of the impact of high population with low levels of affluence and technology. In principle, the existing urban and rural populations might now be better fed with mechanized and more intensive agriculture. But when most farmland is divided into family smallholdings, as it is in Ethiopia, the introduction of large mechanized farms displaces settled communities. This has already happened, amid much publicity, as a result of land being leased to foreign investors.[12]

Food and water

In 1946 the newly-established FAO conducted its first food survey and concluded that at least a third of the world's population would be unable to meet basic food requirements as measured in calories for energy needs. At that time nobody had predicted the long-term outcome of research that had already begun in Mexico.

In the USA, in 1941, President Roosevelt's secretary of agriculture became vice president. His name was Henry Wallace, and before settling into his new office he went to Mexico for the inauguration of President Comacho. It was during that trip that Wallace visited farms, talked to farmers and resolved to improve the sorry state of Mexican agriculture. The upshot of this was a funding agreement by the Rockefeller Foundation to create the Mexican Agricultural Programme in 1943. It was the start of the green revolution.

The first task of the programme was to improve the yields of corn (maize), which was most important in the Mexican diet. It was done by cross-pollinating the best varieties to produce hybrids, then crossing the hybrids. After that, attention turned to wheat. The research was headed by Norman Borlaug, a plant pathologist who in 1970 won the Nobel Peace Prize for his work. In his acceptance speech, Borlaug pointed out that he was only one of a team and that the fight against hunger would not be won until those battling for increased food production were united with those fighting for population control.[13]

It took 20 years of selecting and crossing wheat varieties from around the world but after that time the yields had tripled. The new varieties were more resistant to wheat-rust (a fungal infection) and were more responsive to fertilizers, especially those high in nitrogen. Moreover, they produced heavier heads of seed and had shorter, stronger stalks to support them. The research on wheat showed such promise that the Ford Foundation was encouraged to make its own endowment to agricultural research in collaboration with Rockefeller. The International Rice Research Institute opened in the Philippines in 1962, and other plant research institutes soon followed in other parts of the world. The green revolution was well under way.[14]

There are now 15 research centres working on programmes to benefit the poor of developing countries. They are all financed through the Consortium of International Agricultural Research Centres (CGIAR), a multi-donor trust fund administered by the World Bank as Trustee. On the whole, yield potential, especially for the main cereals, has continued to rise in developing countries. Between 1960 and 2000, wheat yields rose by 208 per cent, rice yields by 109 per cent, and maize yields by 157 per cent.[15] Rates of increase have now levelled off and cereal crops are showing diminishing returns in response to higher inputs.

Despite the indisputable gains in crop yields the green revolution has not been trouble-free and a huge amount of published criticism has accumulated over the years. Some of it is easily dismissed. The crops improved by conventional crossing and hybridization were not controlled by profit-driven corporations because they were never commercial properties. There is no incentive for the private sector to develop products that have no proprietary protection. Seed from CGIAR centres would normally be supplied to national institutions for them to adapt and distribute as appropriate for local conditions.[16]

But it is true that the green revolution largely bypassed the poorest of the poor because the higher outputs from new plant varieties are realized through higher inputs, especially of water and fertilizer, so subsistence farmers who toil for survival in arid regions have not been able to benefit. It is also true that there has been an enormous amount of environmental damage as a result of the higher inputs of water and fertilizer.

About 40 per cent of ice-free land is now devoted to agriculture, not including plantations for wood and timber and land temporarily abandoned in shifting cultivation.[17] Irrigated farms produce 40 per cent of the world's food so agriculture accounts for about 70 per cent of the demand for fresh water; the remainder going into industrial and municipal use.[18] The use of manufactured fertilizer has also soared from around 30 million tonnes in 1960 to an expected 200 million tonnes in 2018.[19] Most of that growth in demand has come from developing countries, mostly in Asia.

Together with the needs of industry, hydroelectric power and domestic use, the agricultural use of fresh water is causing obvious supply problems as well as ecological impacts. Some major rivers, from the Colorado in America to the Yellow River of China, are so overused that for most of the time they peter out before reaching the sea. And as a result of pumping groundwater, the water tables in some places are falling drastically and the depletion of aquifers is on such a scale that it can be detected from space as a result of changes in the earth's gravity.[20] Perhaps half or more of the water stored within a kilometre of the earth's surface comes from pre-Holocene times (in so-called fossil aquifers) and could not be replaced within a human lifetime. Irrigation without proper drainage can also deposit salts near the surface. More than a fifth of the world's irrigated area was salt-affected in 1995, and salinization has got worse since then.[21]

The overuse of fertilizers presents a huge problem. And the combination of fertilizers, pesticides and irrigation causes changes to soil and water so extensive and profound that even a brief outline would require a separate chapter in this book. Because of all the complications and interactions, the broad approach to environmental protection should be to think in terms of ecosystems rather than lists of species and selected pollutants. A European initiative demonstrates this point quite well.

In the year 2000 the European Union passed a Water Framework Directive urging member states to take an ecosystems approach to water quality. The plan was to restore all river basins to 'good ecological quality' by 2015. In the UK it was thought that river quality was not bad. After all, the most persistent pesticides had been banned and there were strict laws to control the discharge of waste into waterways. For the most part, river water was passing the tests for organic pollution and it was getting better all the time. But then, in 2004, the UK Environmental Agency reached the conclusion that more than 95 per cent of British rivers would not meet the new standards.

Under the criteria for good ecological quality, natural biological communities should differ only slightly from those in waters with minimal human impact.[22,23] It turned out that British rivers were by no means exceptional in failing this test; by 2015 nearly half the rivers of Europe had still not reached the new standards.[24]

We can never know exactly how the world would have fared without the green revolution. But in India alone, tens of millions of additional hectares of land would need to have been farmed in an attempt to feed the growing numbers of people.[25] Much of the land would have been unsuited to agriculture, and similar scenarios would have played out elsewhere. So although the green revolution raised the level of environmental degradation, it is possible that the alternatives would have been even worse, both for the environment and for people.

Development

Ethiopia's population policy was introduced at a time when most of the world's governments, after decades of denial or disregard, had come to accept that economic and social development could be adversely affected by population growth. Crucially, governments also managed to agree on the equality of men and women and the need for universal access to family planning. This consensus was reached in 1994 at the International Conference on Population and Development, held in Cairo.[26] A programme of action was approved and, twenty years later, progress was reviewed at a special session of the UN General Assembly. It was concluded that there had been progress but it was unequal and fragmentary.[27]

The United Nations General Assembly of 2000 included a Millennium Summit at which eight Millennium Development Goals were set up to be reached by 2015. First on the list of goals was the eradication of extreme poverty and hunger.

Listed at number seven was the goal of ensuring 'environmental sustainability' but that was not defined. Population growth was not included under any of the goals, and contraception was mentioned only in connection with goal number six—combating HIV/AIDS, malaria and other diseases.[28]

At a summit in September 2015 the UN General Assembly adopted a more ambitious plan of action called *Transforming Our world: the 2030 Agenda for Sustainable Development.* Its aim was to complete, by 2030, what the Millennium Development Goals had failed to achieve during the previous fifteen years.[29] The agenda sets up seventeen 'Sustainable Development Goals' incorporating a total of 169 'targets'. Under goal number five (achieving gender equality and empowering women and girls), one target is to ensure universal access to sexual and reproductive health and rights as agreed in 1994 at the conference in Cairo.

Goals numbered 14 and 15 are directly concerned with the use and conservation of marine and terrestrial ecosystems. Those topics, ecosystems and sustainable development, will be the focus of the next chapter.

19 Measuring up for sustainable development

In a handbook on environmental impact assessment for road-building, the World Bank recounts that a highway was constructed through a tall mangrove forest along a Caribbean coast. The planners had not understood the sensitivity of mangrove trees to the salinity of water. On the seaward side it became too salty, and on the other side there was too much dilution with fresh water. After two or three years the mangroves died, the ecosystem was lost and the local people were deprived of a valued resource.[1] This is just one small example of how human activities can interact with the natural environment in unpredictable ways.

The United Nations, in 2001, initiated a Millennium Ecosystem Assessment (known as the MA) with human well-being as the central concern. The purpose was to assess the global status of major ecosystems and provide a scientific basis for policies to improve conservation and sustainable use. The process took four years and involved more than 1,300 experts in the natural and social sciences from 95 countries.[2] The findings, which comprised five technical volumes and six synthesis reports, were published in 2005.[3]

In summarizing the main findings the overview states that: 'Over the past 50 years, humans have changed ecosystems more rapidly and extensively than in any comparable period of time in human history, largely to meet rapidly growing demands for food, fresh water, timber, fiber and fuel. This has resulted in a substantial and largely irreversible loss in the diversity of life on Earth ...

the degradation of ecosystem services could grow significantly worse during the first half of this century and is a barrier to achieving the Millennium Development Goals.' Nevertheless, it was considered that 'the challenge of reversing the degradation of ecosystems while meeting increasing demands for services can be partially met ... but will involve significant changes in policies, institutions and practices that are not currently under way.'[4]

One observer of the MA admitted that the assessment had proved considerably more difficult than expected and had also revealed the current weakness of ecosystem science. That revelation, together with the severity of the problems, was doubly discouraging to many participants.[5] But I doubt that conservation biologists were greatly surprised by the findings, and anyone who had followed trends in the *Living Planet* reports since 1970 would have been well prepared for shocks.

In describing trends on a broad scale, the threats to biodiversity can be listed under five headings of which habitat loss and degradation would be one. The others, as I mentioned in chapter 15, are invasive species and disease, over-exploitation, pollution, and climate change. These categories of threat are easily distinguished but they can all act in combination, and their impact on a species will depend on the species' abundance, distribution, fecundity, resilience, and whether it is a target for exploitation—legal or otherwise.

For the *Living Planet Index*, trends are revealed by aggregating time-series data for monitored populations of vertebrates around the world. The methodology was developed by the Zoological Society of London and the World Wildlife Fund. In the *Living Planet Report* for 2018 the index is based on over 16,000 populations of 4,000 species. Freshwater types have been hardest-hit, showing an average decline in population abundance of 83 per cent (range 73 to 90% within 95% confidence limits). Overall, the index shows that between 1970 and 2014 there was an average decline of 60 per cent (range 50 to 67%) in population sizes of all vertebrates (mammals, birds, reptiles, amphibians, fish).[6]

Ideally, there would be a series of indices, going back, say, to 1950, showing the changing composition of hundreds of ecosystems representing the world's major 'habitat types'.[7] Nothing remotely like that exists, nor is there the level of understanding to interpret all that data if it could magically be made available. Scientists can only make guesstimates as to the total number of species, and very few ecosystems have been studied to the point of finding out how healthy functioning is maintained. The *Living Planet Index* is an impressive achievement but population counts of vertebrates provide only a first indication of what is happening to the natural world.

The ecosystem level of ecology is key to understanding how the natural environment supports humans as well as other forms of life. Nevertheless, some species do act as useful indicators. For example, it is reasonable to assume that if tigers are doing well in a protected forest, and nobody is losing livestock, then deer and other wild prey are also doing well, as are their food plants and the soil that supports them. Healthy tiger populations indicate a healthy forest. If the tigers were exterminated there would be a cascade of changes through the food webs of the forest that could be hard to predict in detail. On the other hand, if soil fertility declined because soil organisms were dying out then we might not be aware of it until changes began to show up in the forest vegetation. But that would be alarming because the cause and consequences could be widespread and even harder to fathom.

Species that are easily seen and counted might prove to be useful indicators of healthy ecosystems. Some might possibly serve, like the canary in the coal mine, as environmental sentinels for people, but ultimately, the health of all ecosystems depends on the workings of obscure little organisms and the cycling of nutrients. I make this rather obvious point because of its relevance to the concept of sustainable development, which we will reach after the next section.

Ecosystem services

The range of attitudes and feelings towards man's place in nature has not diminished but since the 1960s there has been a surge of interest in establishing the links between ecosystems and their economic benefits to human well-being.[8] Some of those who valued nature for its own sake were pleased to have their views supported by news that the benefits from natural ecosystems were worth a fortune in dollar terms. Others, who were interested only in the utility value of nature, needed no convincing that the economic approach was the right one.

Of course there are some who still insist that nature is priceless and cannot be reduced to a commodity with a cash value. Many proponents of the shift to economic thinking do not disagree with that, at least not in principle, but they see it as the nub of the conservation problem: when natural goods and benefits are said to be priceless they are given no monetary value and their loss cannot show up in the national accounts.[9] With no entry in the accounts the priceless can be regarded as worthless. Conservation biologists therefore joined the bandwagon and rejoiced in trying to find the dollar value of ecosystems. That has also attracted criticism from among those with a strictly utilitarian point of view. They protest that if the conservationists' agenda is really to protect biodiversity for its own sake, then the valuation assessments should be seen as 'advocacy tools' and are not 'scientific exercises'.[10] I cannot see why the use of scientific findings by advocacy groups necessarily invalidates the science.

In deciding whether nature really does have its own intrinsic value we must very carefully define our terms before plunging into the morass of ethical, philosophical and moral arguments. So far, nobody has emerged with a definitive answer so the arguments will continue among those who are not content with mere opinions. For the record, the MA Biodiversity Synthesis simply asserts that 'biodiversity and ecosystems also have intrinsic value.'[11] In any case, intrinsic value is not the only alternative to cash value; nor are different kinds of value mutually exclusive. Most people would probably agree that living nature deserves a kind of respect that would not be appropriate for sand and cement. And if it is difficult, ethically and philosophically, to prove that nature has intrinsic worth, it is equally difficult to show why benefits to humans are all that should count towards a thing's value.[12]

The importance of putting a value on 'nature's services', and the difficulty of the task, were first considered in a science journal in 1977.[13] In the decades that followed there were scores of attempts to attach price tags to what had become known as 'ecosystem services'. In 1997 one team of scientists synthesized the findings from over a hundred of the studies and came up with an average of US$33 trillion (range US$16–54 trillion) for the value of the entire biosphere. At that time the global GDP was about US$18 trillion. Acknowledging the uncertainties, the authors of that milestone paper considered their estimate to be a minimum and they emphasized that it was only a starting point for future research.[14] It may well have helped to trigger the avalanche of paperwork that followed.

The Millennium Assessment classified ecosystem services into four very broad types under the headings of provisioning (of natural products, cultivated crops etc.), regulating (of climate, floods, clean water and other needs) cultural (including aesthetic, recreational, spiritual and educational purposes) and supporting. This last group of services included processes such as nutrient cycling and soil formation that served to underpin the other three.

Services within the cultural group are particularly intangible and difficult to represent with a dollar sign. Economists can record costs actually incurred by those individuals who spend money in using cultural services, and it can also be useful to record the cost of restoring losses to services. In other cases, contingent valuations must be obtained through methods such as using questionnaires to reveal willingness to pay for preventing losses, or amounts acceptable as compensation for losses that have occurred. Contingent valuations are obviously far from ideal. There are also shortcomings in the MA classification scheme itself. The few basic categories may have been enough for the initial, descriptive phase of assessment but they are not adequate for valuation purposes.

Semantic uncertainties first need to be settled so that words such as 'service', 'process', 'function' and 'benefit' have workable definitions even if precision is impracticable.[15] The word 'value' must be clearly understood if we are considering different types of value. Then there are overlaps to consider and double counting to be avoided. For example, supporting and regulating services will contribute to the value of wood from a forest so those contributions should not be separately valued in addition to the wood itself. The spatial boundaries of the ecosystem being valued need to be defined and the beneficiaries need to be identified. And where ecosystems have been deliberately modified to increase productivity, the human inputs must also be taken into account.

One attempt to improve the classification system began in 2008 with support from the European Environment Agency. The outcome, called the Common International Classification of Ecosystem Services (CICES), is the scheme now used by the European Union in pursuing its own aim of halting the loss of biodiversity and ecosystem services in member countries by 2020.[16] Other classification systems have been created using CICES as a basis and there are likely to be many refinements and adaptations in the years to come; these are early days in the endeavour.[17,18]

In its current version, CICES has a hierarchical structure with five levels designed to avoid overlaps. At the highest level are four Sections. These correspond to the broad MA categories except that 'supporting' has been replaced with 'regulation and maintenance'. Nested below the Sections, and becoming progressively more specific, are Divisions, Groups, Classes and Class types. The creators of the system suggest that Class types could be useful in mapping and assessing but the Class level should be appropriate for accounting. For example, 'cultivated crops' would be a Class, whereas 'cereals' would be a Class type.

An incredibly complex world has resulted from natural selection and human activity, and the challenge in any valuation approach is to schematize that world and reduce it to the outputs that directly contribute to human well-being. There is no point in having a classification system more complicated than it needs to be, but ecosystem outputs that happen to be of direct benefit to human well-being are not self-supporting; they result from the awesome complexity that is not revealed by a hierarchy of ecosystem services. Some critics see this as a deficiency of CICES and similar schemes.[19]

Where and when losses of ecosystem services are counted as monetary losses they will be gathered together with other costs in the national accounts and will affect a nation's GDP. Otherwise, environmental disasters can boost economic activity and so will count as a plus in compiling GDP. In 1989, for instance, when the oil tanker, *Exxon Valdez*, crashed onto the Alaskan shore, the resultant oil spill cost more than $2 billion to clean up. Spending that amount would have added to the year's GDP, with economic downturns in tourism and fishing offsetting the gain in subsequent years. It is widely believed that, on balance, the *Exxon Valdez* disaster was a gain for America's GDP but data from different sources are aggregated in the accounts and cannot be separately identified.[20]

The event was particularly significant in the present context because contingent valuations were used in calculating the amount of compensation that Exxon was required to pay. When Exxon tried to discredit the methodology, the US Department of Commerce appointed a panel of independent experts to look into it. The panel supported the valuation, and the method, concluding that it could produce estimates reliable enough to be the starting point of a judicial process of damage assessment.[21]

Every nation is free to develop its own system of national accounts (SNA) but since 1953 there has been an internationally recognized framework that was last revised in 2008. That revision, mandated by the UN Statistical Commission, was produced by five international organizations and is intended to serve as a guide.[22] However, the framework does not address the need to incorporate environmental statistics in accordance with recommendations from the Brundtland Commission, the Rio conference, and the UN Conference on Sustainable Development (Rio+20).

A handbook on integrating environmental and economic accounting was developed in 1993 and in its current revision it is available as the *System of Environmental-Economic Accounting 2012—Central Framework* (SEEA Central Framework).[23] The framework shows how to deal with the stocks and flows of environmental assets but falls short of providing guidance on how to include ecosystem values, as distinct from known environmental statistics, in national accounts. That is not to suggest that the problems were overlooked. Alongside the SEEA *Central Framework*, another document was approved by the UN Statistical Commission—namely the *SEEA-Experimental Ecosystem Accounting*.

The *SEEA-Experimental Ecosystem Accounting* is complementary to the *Central Framework*, but as its title suggests, it is not an international standard so much as 'an important first step in the development of a statistical framework for ecosystem accounting'. The expectation (or hope) is that countries will test and experiment with it in the coming years.[24]

Sustainable development

The idea of sustainable development has become inordinately popular but just what it means in practice is hard to pin down. As I said in chapter 15, the definition most commonly encountered is based on an extract from the Brundtland Report of 1987. It calls for development to meet the needs of the present without compromising the ability of future generations to meet their own needs. That might sound like a reasonable guiding principal but it is not one that can be scrupulously followed.

Some essential development goals will have unavoidable environmental impacts. In *Transforming Our World*, goal number 9, for instance, includes building resilient infrastructure and promoting 'sustainable industrialization'. How, exactly, is that to be interpreted? The problems of finding criteria for sustainability were not addressed in the outcome document of the 2015 summit. Instead, an expert group was set up to develop, within a year, a framework of indicators for monitoring progress in meeting the goals and targets.

The expert group duly put forward one or more indicators for each of the 169 targets. For example, one indicator for promoting sustainable industrialization is the extent of manufacturing employment as a proportion of total employment. An indicator for building resilient infrastructure is the proportion of rural populations living within 2 km of an all-season road. Criteria for assessing sustainability remain as elusive as ever.

As for the needs of future generations, all we can say with any certainty is that basic human needs for the remainder of this century will continue to increase as they have done in the recent past. In 1985, when the Brundtland Report was being put together, the per capita GDP of the OECD countries was US$12,561. It is now nearly three and a half times that amount.[25] Over the same period the world population has grown by 2.7 billion.

So there is no strict recipe, prescription or formula for sustainable development. Case by case, it will be a matter of making judgements about the gains and losses, the trade-offs, involved in new development proposals. The intention is to enable judgements to be made in the light of ecosystem valuations as described above, and in consideration of future generations as well as present beneficiaries. Progress so far has not been at all encouraging. In 2016 The European Commission's senior adviser on sustainable development described it as 'a concept poorly implemented, if at all'.[26]

If tools for ecosystem valuation become universally acceptable, and criteria for assessing sustainability are developed, we shall have reached the stage where it will remain only for the world's decision-makers to use them to good effect. That could prove to be the biggest challenge of all. In this context it seems appropriate to close the chapter with two or three paragraphs about a high-profile development decision that is currently raising passions in Australia.

The proposed Carmichael coal mine in Central Queensland will be one of the biggest coal mines in the southern hemisphere. The plan is for 6 open-cut pits, 5 underground mines and a 189 kilometre rail-line to ports on the Queensland coast from where the coal will be loaded for shipment, mostly to India. The proponents of the mine claim that it will create up to about 2,500 jobs in the construction phase and possibly close to 4,000 jobs when fully operational.[27]

Lying off the Queensland coast is the Great Barrier Reef, the world's largest system of coral. According to Deloitte Access

Economics the reef has an economic, social, and icon asset value of US$42.4 billion (A$56 billion) and it contributed A$6.4 billion to the Australian economy in 2015–16.[28] The reef is also a UNESCO World Heritage site and is carefully monitored for the long-term effects of climate change, coastal development, declining water quality and other threats.[29] During the past two years the reef has suffered widespread coral decline, mainly as a result of above-average sea surface temperatures.

Predictably, environmental groups are solidly opposed to the Carmichael coal project but both state and Federal governments are in favour of it. And the State government has particularly angered local farmers by granting the proposed mine unlimited access to groundwater for sixty years. If India buys its coal from another source, the governments contend, it will probably be lower quality coal and will result in even higher carbon emissions. It seems likely that the availability of finance will be the decider.

Nobody said that sustainable development would be easy, but for as long as 'development' is used as another name for 'growth' there will be reason to question whether it can ever be more than an idea, and an ill-defined one at that.

20 Growth

Historically, economies have grown along with human populations because we all need to buy at least the basics of life, but as we have seen in earlier chapters, events in the nineteenth century resulted in some nations leading the way to higher productivity so that economies around the world began to outstrip population growth. That is what matters to individuals because if economic growth merely keeps pace with growing populations then there is nothing extra to share and people who take bigger slices of the economic pie must be taking away from others. Politicians typically fail to mention this obvious fact and announce strong national growth in GDP as if it calls for unreserved applause. They also tend to make misleading comparisons.

In 2016, for instance, it was proudly announced that Australia had beaten the record held by the Netherlands for the longest period of uninterrupted growth in GDP. What was not widely mentioned was the fact that Australia's per capita growth was poor compared with that of the Netherlands. Australia's high immigration programme inflated aggregate GDP while producing no increase per head of population.[1] During 2016 Australia's population grew by around 1.6 per cent, compared with 0.32 per cent in the Netherlands. Australians who looked at the growth of their economic pie-slices would not have seen so much to feel cheerful about.

Not only that. In recent years many people in Australia have begun to question the wisdom of promoting strong population growth as they battle worsening traffic congestion, longer commuting distances, unaffordable housing and a stubbornly high rate of unemployment.[2] The questioning minority is making itself heard but the response from most economists and the main political parties is almost always entirely predictable: strong population growth is to be welcomed so there is an urgent need for more road and rail links, better public transport, more job creation and the approval of more land for housing and associated development. These are all measures guaranteed to raise GDP and increase the destruction of the natural environment. It is a scenario that will be familiar to people in many countries around the world, and it bundles together a number of questions about growth that will occupy most of this chapter.

Economic growth per person is a very recent phenomenon. True, there were surges of development in early times but for a thousand years there was no sustained growth in per capita GDP; economies merely expanded along with populations. In the second millennium there was very slight per capita growth until about 1870, after which it averaged around one per cent a year until the approach of the golden age of economics in 1950.[3] But although economic growth has been sustained since then it would be wrong to think of steady progress; there were global recessions in 1975, 1982, 1991 and 2009. And between 1970 and 2013 the nations of the world were affected by more than 400 currency, banking, and debt crises.[4] Moreover, there have been huge changes in the make-up of the global economy over recent decades.

Broadly speaking, in the early 1970s the nations of the south were poorer than those of the north and supplied raw materials and agricultural produce. Manufactured products from the north made up most of the global trade. Since that time some rapidly-developing nations, with China prominent among them, have expanded their economies to such an extent that they now account for more than 75 per cent of global growth in output and consumption. China has become the world's second largest economy in terms of nominal GDP, with India and Brazil in 6th and 8th place respectively. This catch-up of 'emerging market and developing economies', as the IMF calls them, is expected to continue into the immediate future.[5,6]

Theories of growth

In the longer term there is not much economic certainty, but without doubt the babies of today will reach old age in a very different world from the present one, and economics will contribute to the transformation. Economic growth, which barely existed one and a half centuries ago, is now pursued relentlessly by all governments as well as economic organizations with international influence. In the future, as in the recent past, this will bring about profound changes that are not always intended, foreseen or even understood.

We have quite good records of economic progress, setbacks and consequences since economies first began to grow in a sustained way but how did they get going to begin with? What was it in the nineteenth century that made some populations more productive in relation to their numbers and first led to sustained economic growth? The short answer is that nobody really knows. Nearly a dozen possibilities have been suggested for the UK alone but specialists admit that it could be any, all, or none of them.[7]

It is not merely a question of academic interest. In the history of economic growth, scholars have hoped to find answers to matters of practical importance. For instance, after sustained growth took off, why did it diffuse to some countries but not others? And, once it became established, how did it remain so resilient despite serious disruptions? In cases where the details are best known there is no common pattern to support an explanation, and even our most basic understandings have changed over time.

In the thinking of Adam Smith, David Ricardo and others of the classical school, the costs of production and distribution of an object were what determined its market value—it is one of the tenets of classical economics. But in a much-quoted example, a person dying of thirst would surely pay more for water than for diamonds, regardless of the costs of having made them available. People in real life have personal reasons for assessing the worth of something, and as we all know, one person's trash might be a treasure for somebody else.

By the 1870s a number of economists had accepted that 'subjective elements' were involved in setting market value so it seemed more realistic to think in terms of supply and demand rather than just focusing on production and distribution costs. Utility, in the sense of consumer satisfaction, became the measure of importance in what would become known as neoclassical economics. But it was still assumed that people behave rationally, which is to say that firms make decisions that maximize their profits, and consumers make decisions that maximize utility.

It sounds reasonable but it has been amply demonstrated that we don't just make calculated decisions in everyday life, we also react and respond intuitively and spontaneously.[8] We interpret facial expressions and tones of voice without being aware of any thinking process at all. Similarly, we cannot always predict the moods and feelings that influence us, nor always explain them if we give them some thought. To the extent that there is survival value in functioning this way, it must have proved its worth in the Pleistocene epoch during our formative years as a species. We have acquired an enormous amount of knowledge since economic growth began but our brains have not been fundamentally restructured.

What has changed during recent centuries is the marketplace. When economic growth first took off, the world of advertising blitzes, sky-cluttering air transport, smart phones and transglobal chatter would have been hard to imagine. All things considered, it is not really surprising that a high priest of economic growth theory recently declared that 'for the last two decades, growth theory has made no scientific progress towards a consensus.'[9]

The need for growth

In the previous chapter I referred to the Sustainable Development Goals that were set up by the UN with targets to be achieved by 2030. The first of the seventeen goals is to eliminate poverty. Goal number eight is to 'promote sustained, inclusive and sustainable economic growth, full and productive employment and decent work for all'. One of the targets of that goal is to achieve at least seven per cent GDP growth per year in the least developed countries.

The least developed countries (LDCs) are identified by the UN on the basis of income, human assets (health and education) and economic vulnerability.[10] The list is reviewed every three years, and at the last count (2016) there were 48 of them, of which 34 were in Africa.[11] Almost half the total population of these countries lives in extreme poverty, which at the personal level means surviving on no more than US$1.90 a day.[12] With so little money to spend, villagers with access to water and enough land to grow vegetables and keep chickens could have a much healthier life than those who live in city slums.

The aggregate GDP growth for the least developed countries was 4.5 per cent in 2016, with the expectation of reaching 5.5 per cent in 2018. Growth rates vary significantly between the LDCs, however, and some of them, including Ethiopia, Bangladesh, Rwanda and Myanmar, could reach the seven per cent target in 2018.[13] In fact, the growth rates of the LDCs are presently comparable to that of Asia as a whole and are significantly higher than those of the advanced economies, which average about two per cent.[14] The emerging market and developing economies are therefore catching up, though different measures can show different trends as I will explain in the next section. Since the 2008 financial crisis, more than 80 per cent of global growth has been generated by emerging market and developing economies, and their share of world GDP is not much below 60 per cent, which is nearly double what it was in the mid 1980s.[15,16]

This has brought about some progress in reducing poverty. In 1990, according to the World Bank, 35 per cent of the world's people lived in extreme poverty and by 2013 the proportion had been reduced to 10.7 per cent. This does not mean that numbers had been reduced to less than a third; population growth during that period must be taken into account. Even so, if the figures are correct, the number of extremely impoverished people in the world was more than halved, with the number falling from 1.84 billion in 1990 to 767 million in 2013.[17] While that is good news, a total of 767 million is still far too many—it is actually one and a half times the population of the European Union. It is also recorded that during the 1990s there was a particularly high number of people just below the extreme poverty line, so in large part, the successes did nothing more than swell the ranks of those immediately above it.[18] There is obviously a long way to go.

And the needs are daunting. The International Labour Organization estimates that about 40 million new jobs will be required every year until 2030 just to keep pace with the growth of the working age population.[19] As things are, nearly half the workers of southern Asia and nearly two thirds of the workers in sub-Saharan Africa are earning less than US$3.10 per day in terms of local purchasing power. More than 42 per cent of employed people (1.4 billion) are in vulnerable or precarious forms of employment, and more than 200 million are unemployed.[20]

With facts like this we must expect governments to be doing whatever they see fit in an effort to boost growth. Robert Wade, at the London School of Economics, describes economic growth as 'the great engine of escape from material deprivation'.[21] Mainstream economists of all persuasions would surely agree with that but to some extent they would also share Professor Wade's view that it is not enough. To argue that economic growth alone will serve to put money where it is most needed requires an unshakable faith in the virtues of the free market. Nowadays all countries apply some mix of interventions under the heading of social policy; the details depend on national politics and opportunities, and in western-style democracies there is no shortage of columnists and bloggers to supply the mass media with different opinions. I shall be content to summarize a few more global facts.

Inequality

Since temples became the first bases of ruling elites there have been distinct divisions through societies, and like everything else in civilization, the divisions have become a lot more complicated over time. One popular measure of income inequality is the Gini coefficient, named after the Italian statistician, Corrado Gini, who introduced it in 1912. It is a measure of distribution so its use is not confined to income data. The scale runs from zero to one. A Gini of zero indicates perfect equality in which everybody receives the same income, and a Gini of one would result from all the income going to one household, so in reality, Gini coefficients always work out at something in-between. Multiplied by 100 they can be expressed as percentages, but when they are based entirely on income distributions they offer no clues as to actual incomes or size of economy. In 2015, for example, the USA and Turkey both had a reported Gini coefficient of 0.39, or 39.

Economists have standard definitions of income to include such niceties as imputed values for home-grown food, but they are unlikely to have been consistently applied in household surveys, and Ginis are no more accurate than the survey data on which they are based. In general, higher incomes tend to be under-reported. In a study by the Brookings Institute the discrepancies between household survey data and national accounts figures were used to rework more than a thousand Gini coefficients of 134 countries for the previous 25 years. This adjustment lifted the average within-country Gini from 39 to 48—an increase of 9 percentage points, though there was a considerable spread around the average. Some countries were unchanged whereas Nigeria shot up to become one of the most unequal countries in the world.[22]

Like many single-number measures, the Gini coefficient may not be the right one to tell us what we want to know. For example, the ratio of two thousand to twenty thousand dollars is the same as that of four thousand to forty thousand dollars so the Gini would show the same degree of inequality. Yet the difference in pay-packets has doubled. An analysis based on absolute values would show a different picture from that based on ratios. As Robert Wade concludes, there is no best measure of world income inequality. Different measures are useful for answering different questions.[23]

Branko Milanovic is an acknowledged authority on the subject. In 2012, while he was lead economist with a World Bank research unit, he produced a working paper on global inequality with particular attention to the period between 1988 and 2008—which he considered to be the most recent episode of globalization. Earlier periods could not be compared with much precision because too few countries had conducted household surveys involving actual incomes.[24]

Among other things, Milanovic's paper clearly shows the different pictures that emerge when we compare nations for inequality, with and without population sizes being included. Plotting Gini coefficients for average incomes, and ignoring population size, results in a graph showing the Ginis growing steadily higher between about 1980 and the turn of the century. Nations were becoming more unequal for income. But when the Ginis are calculated for average incomes with population size included we see the opposite trend; the Ginis grow smaller to show the countries converging as average incomes become more equal. Naturally, the proponents of globalization will point to one graph and critics will point to the other. The apparent disparity is accounted for by the rise of the Asian (mainly Chinese) middle class when their great numbers are taken into account.

Currencies are made comparable by adjusting for purchasing power within countries, but if actual incomes are used, rather than country averages, the Ginis will always be higher because averaging-out lowers measured inequality.[25] Using incomes, the global Gini for 2011 worked out at about 70, which would mean that if the world's income were divided into two halves, eight per cent of the population would take one half and the other 92 per cent of the population would share the other half. The equivalent numbers for the USA would be 22 and 78.[26] Income inequality between countries is higher than that within any single one, although a few countries, including South Africa and Haiti, have Ginis of over 60.[27]

The analysis by Milanovic supported the common assumption that the winners from two decades of globalization (1988–2008) were the rising middle classes of the emerging market economies, and the top one per cent of earners globally. But other changes were more surprising. While the top one per cent made great gains, the next five per cent also did well, but the twenty per cent below them gained little or nothing in real incomes. Neither did the poorest five per cent make any headway. With outcomes like that, arguments about the pros and cons of globalization will not be easily settled.

People with higher incomes also tend to have more wealth but it is not always the case. The annual *Credit Suisse Global Wealth* reports are a valuable and respected source of information on the distribution of wealth, and for the remainder of this section I shall use data from the most recent edition.[28] When wealth is defined as total assets (financial and material) minus debt, it is easy to see that wealth can be determined by debt for individuals who have very different standards of living. Some may have just bought a luxury home and its value has slipped below the amount of outstanding mortgage. Others may have borrowed heavily to buy a goat or a hand-operated water pump.

It is estimated that nine per cent of the world's adults are net debtors, which is to say that their debts exceed their assets, including property. I have no idea whether that will come as a surprise to readers, but I feel sure that many will be aware that the richest one per cent of households has slightly more than half the world's wealth. It would have been hard to avoid the publicity attracted by that piece of information. It is a shocking degree of inequality, and it has crept up slightly since the turn of the century.

The true measure of it is more sobering because one does not need to be super-rich to be in the top one per cent: the sum of US$744,400 would get you there. Any family having paid the mortgage on a house in an expensive city would probably be included. Half of all adults in the world, that is about 2.4 billion men and women, possess less than US$2,222 in total wealth, and only 27 per cent of adults in the world have more than US$10,000. That includes all financial and non-financial assets, minus debt. Regional differences are pronounced. For example, whereas the residents of Latin America range fairly evenly across the wealth spectrum, African nations and India have extreme wealth inequality, with large proportions in the category of lowest wealth. In these countries the share of global wealth is less than a tenth of their share of global population. In contrast, China has relatively few people in the lower half of global wealth distribution, with most being in the upper-middle section.

And yet the number of millionaires in Africa and India has grown faster than the global average this century, and the number in China has grown spectacularly from no more than 45,000 to about 1.6 million. The USA has the most millionaires and individuals of 'ultra-high net worth' but China holds second place with 11,000 individuals being worth more than US$ fifty million. In terms of wealth, Russia is the most unequal of the major economic powers, with the richest ten per cent owning an estimated 89 per cent of all household wealth. The corresponding figure for the USA is 78 per cent, and for China it is 73 per cent.

Perhaps the most publicised finding of all is the wealth of the richest individuals. In January, 2017, Oxfam announced that the wealthiest eight billionaires, headed by Bill Gates at that time, controlled about as much wealth as the poorest half of the world's population. The calculation, which included net debtors, was based on data from the Credit Suisse report and the *Forbes Billionaires List* which can change daily along with fluctuations in stock prices and exchange rates. According to Oxfam, if net debtors are excluded from the estimates, the wealth of the poorest half would be roughly the equivalent of the richest 56 individuals.[29]

The Oxfam report was released in time for the World Economic Forum that is held each year in Davos, Switzerland. The Forum is a not-for-profit foundation that was established in 1971 with headquarters in Geneva. Its mission is to improve the state of the world by encouraging the cooperation of public and private sectors, international organizations and academic institutions.

At the meeting in 2017, a time of rising populist movements, increasing nationalism and anti-globilization, the topic of inequality was unusually high on the agenda.

Many delegates agreed with Christine LaGarde, managing director of the IMF, when she called for policies to redistribute wealth in a fairer way. For most delegates that would be interpreted as tax reform but there were others in favour of income top-ups from governments in order to maintain a universal basic income. Yet there were also some who argued in opposition, believing that growing a bigger pie was a higher priority than working out how it should be shared. After all, the more money coming in, the more there is to trickle down and spread around. Delegates must have heard it all before. International organizations can have a lot of influence but, ultimately, policy changes must be implemented by national governments.

On the subject of spreading it around, it is worth bearing in mind that if all the wealth of those eight richest men (about US$426 billion at the time) had been shared equally among the world's adults (then numbering 4.8 billion), they would each have received about US$89. While that would be a wonderful gift for some individuals, it would scarcely affect the global needs outlined above.

Limits to growth

The idea of a population being able to grow eternally on a single planet belongs to the realm of science fiction so the question comes down to whether there are limits to economic growth while populations are stable or declining. As I said in chapter 14, the demands created by the rate of growth in the 1960s was deeply troubling to those with a concern for the natural environment, and it was alarming to many who believed it to be unsustainable. At that time population growth and economic growth were usually conflated in the arguments that flared through the opinion columns.

In 1968 a group of thirty individuals from ten countries met in Rome to consider the gravity of the situation. At the meeting were scientists, economists, industrialists and civil servants, both national and international. It was this gathering that gave rise to the the Club of Rome, now a global think tank that limits itself to around 100 active members with a secretariat in Switzerland.[30] As one of its first initiatives, the Club commissioned a team of researchers, based at the Massachusetts Institute of Technology, to develop computer models of different global development scenarios. Using system dynamics theory and a computer model called 'World3', the team analysed 12 possible scenarios and environmental outcomes to the year 2100. The modelling showed that the 'business as usual' scenario was unsustainable, and the Club of Rome released a non-technical version of the findings as its first report. It was published in 1972 with the title *The Limits to Growth* and it sold twelve million copies in more than thirty languages.[31]

There was already widespread public concern about toxic pollutants because Rachel Carson's book, *Silent Spring*, had been published ten years earlier. Then, in 1968, Paul Ehrlich's book *The Population Bomb*[32] had warned of imminent famine, and now *Limits* was sounding additional alarms about the depletion of essential resources and a level of CO_2 in the atmosphere that was predicted to reach 380 ppm by the year 2000 (it was reached in 2005). Environmentalists largely welcomed the publications though some on the left of politics were critical that neither *Limits* nor the *Bomb* had addressed the issues of social justice. They took the view that the root of the problem was over-consumption by societies that were already rich, so they failed to see the importance of human numbers and growth itself. On the other side of politics, some proponents of fast growth and big business were just as passionate in denouncing the books as nothing more than irresponsible pseudo-science. Critics of that sort are still vocal.

A more thoughtful proponent of big business and unbridled growth was the late Julian Simon (1932–1998), an optimistic economist if ever there was one. Simon was so strongly in favour of population growth that he seemed to be driven by religious, rather than economic, fervour. We get some insight into his philosophy from his assertion that: 'enabling a potential human being to come into life and to enjoy life is a good thing, just as protecting a living person's life from being ended is a good thing.'[33] He saw no economic problem because he believed that 'our supplies of natural resources are not finite in any economic sense.' Simon was confident that alternative materials and improved technology would sustain growth indefinitely and standards of living would continue to improve. Even agricultural land, Simon wrote, 'is not a fixed resource ... and is likely to continue to increase where needed'. Moreover, 'the long-run impact of additional people is likely to speed the development of cheap energy supplies that are almost inexhaustible.'[34]

But the 'most important benefit of population size and growth', Simon believed, 'is the increase it brings to the stock of useful knowledge'. This is the argument that more people will produce more clever ideas to contribute to progress. Simon doesn't say why people with good ideas are more likely to crop up among the millions than those with bad ideas. For what it's worth, when we look at the number of Nobel prizes awarded per million people, the smaller countries have been the most successful.[36] Herman Daly, whom we shall meet in the next section, declared that Julian Simon made mistakes faster than his critics could correct them.[37]

Even so, Simon's irrepressible optimism may have swayed public opinion quite significantly. In 1980, he famously made a bet with Paul Ehrlich that a basket of five metals (chromium, copper, nickel, tin, tungsten) would fall in price between 1980 and 1990, instead of becoming more expensive as one might expect with increasing demand. Simon won the bet.[38] Furthermore, Ehrlich's predicted famines did not materialize. It is particularly risky to make public predictions with a time-limit. If the forecast event does not happen, then the attentive public might give a mental shrug and promptly lose interest as if the whole issue has gone away with the passing of the deadline. There may well have been a waning of interest in the subject but the authors of *Limits to Growth* undertook a follow-up study, thirty years later, from which they concluded that despite new technologies, new institutions and a new awareness of environmental problems, there was far more cause for pessimism than there had been in 1972. More recent studies support that conclusion.[39,40]

The World Bank, having decided by 2006 that the benefits of growth were not fully appreciated and the causes of growth were not fully understood, set up a Commission on Growth and Development. Two years later the Commission released its *Growth Report*, reaffirming the view that growth is a necessary, if not sufficient, condition for broader development, and that 'developing countries cannot grow without the support of the advanced economies. In particular, they need access to the open global trading system.'[41] On the question of whether growth has limits, the report says, 'the answer will depend on our ingenuity and technology, on finding new ways to create goods and services that people value on a finite foundation of natural resources.' At least the Commission acknowledged that natural resources are finite, but only climate change was considered to be a problem.

Decoupling

The word 'decoupling' in the context of growth, seems unlikely ever to rival 'sustainability' as a buzzword, but the two have become firmly linked in the thinking of some international organizations. In principle, the idea is that economies can keep growing without doing unacceptable damage to the natural environment because it is possible for environmental pressures to be decoupled from economic growth. The OECD environment ministers adopted decoupling as one of the main objectives of the OECD Environmental Strategy in 2001.[42]

It is not an entirely new challenge; economies have been trying to do more with less for a long time in the effort to become more efficient. For example, it is reported that in 1969 a dollar's worth of GNP was produced with half the materials that would have been needed for the equivalent production in 1900. This shows a relative decoupling because less material was used up for the same amount of growth, but nonetheless, more growth required more material. Over that same period total material consumption increased by 400 per cent.[43,44] In order to bring about an absolute decoupling there would need to be continued economic growth with zero increase in extraction of materials, and as the use of material resources is only one kind of environmental impact the same would need to be done for all other kinds as well. Only then could we say that economic growth and environmental impact had been completely decoupled. On the face of it, it sounds impossible.

Europe, with advanced economies, mature institutions and relatively stable or slowly declining populations, can be expected to have performed as well as anywhere in attempting to reduce environmental impacts so it is reasonable to look there for signs of progress. The signs there will also be easier to see because the EU is attempting to monitor progress by maintaining a resource efficiency scoreboard. On the scoreboard a number of indicators are arranged in a hierarchy headed by 'resource productivity', which is defined as GDP divided by domestic material consumption (DMC). Below that are macro-indicators for materials, land, water and carbon. Each of these has a number of specific indicators (32 in all). This is a pioneering attempt to measure resource efficiency by a number of criteria on a large scale.[45]

So, for example, between 2002 and 2014, resource productivity improved from 1.52 euros/kg to 1.95 euros/kg, and the use of materials per capita fell by over 13 per cent, though there was wide variation across Europe. However, the use of DMC can be misleading because it doesn't take into account the energy and materials embodied in the imports of manufactured goods from outside Europe. Eventually DMC will be replaced by raw material consumption (RMC). With regard to waste disposal, the recycling rate for municipal waste increased from 25 per cent in 2000 to 42 per cent in 2013, with a corresponding reduction in landfill. The target is to move towards a circular economy in which nothing is wasted. Also on the plus side there was an overall downward trend in greenhouse gas emissions per person after 2000, but it was still far from being on target.

On the minus side, more river basins are being affected by water stress and that trend is expected to increase by up to 50 per cent by 2030. The loss of land to infrastructure (roads, parking areas, buildings) also appears to be increasing. The rate of loss between 2000 and 2006, if continued, would result in an area the size of Cyprus being paved over every ten years. Preliminary findings suggest that the rate was higher during the subsequent three-year period. Habitats are becoming more fragmented as a consequence of continuing 'land-take', and this can reduce biodiversity as described in chapter 15. The EU scoreboard uses a count of 167 common bird species as an indicator for trends in biodiversity. Overall, numbers declined by 16 per cent between 1980 and 2013.[46]

In summary, Europe has made progress wherever technology can be applied to using materials more efficiently and with less associated damage to the environment, but land-use conflicts have not been reduced and there is still a growing impact on living nature. There is no evidence of absolute decoupling, and in terms of resource efficiency on a global scale, 'even relative decoupling just isn't happening'.[47]

Alternative economies

John Stuart Mill, writing in 1848, could see no good reason why economic growth should continue indefinitely. As he saw it, 'the best state for human nature is that in which, while no one is poor, no one desires to be richer, nor has any reason to fear being thrust back by the efforts of others to push themselves forward.' Nor could he see why it should be 'a matter of congratulation that persons who are already richer than any one needs to be, should have doubled their means of consuming things which give little or no pleasure except as representative of wealth'.[48] In Mill's view of the world, when economic growth had done its good work, there could be a 'stationary condition of capital and population' in which there would be 'as much scope as ever for all kinds of mental culture, and moral and social progress'. Many people today have a similar vision for a better future but on some measures reality is moving further away from it and the direction of travel seems firmly set.

Present-day notions of a steady state economy owe much to the pioneering work of Herman Daly, who has been writing on the subject since he edited an anthology on it in 1973.[49] Daly, currently emeritus professor at the University of Maryland, was a senior economist in the Environment Department of the World Bank, and was a co-founder of the journal *Ecological Economics* in the 1980s. He set up the Centre for the Advancement of Steady State Economics (CASSE) in 2004 to pursue the goal of a steady state economy through education, promotion and study.[50]

Basically, whereas J. S. Mill seemed to think that economies might eventually settle into a 'stationary condition', Herman Daly saw that if modern societies are to live within the natural cycles of the biosphere then governments will need to ensure that it happens because it will not happen by itself and the laws of nature will not make exceptions. Basically, a steady state economy would aim for stable or mildly fluctuating levels in population as well as in the consumption of energy and materials. Production rates would equal depreciation rates. There is now a substantial literature on the steady state economy but most of it is concerned with promoting it as an alternative to the destructive growth imperative that exists. Promotion is not difficult to do. The hard part is to find an acceptable way of moving from one to the other; acceptable, that is, to those in positions of enough power and responsibility to bring about such change, should they become convinced of the need for it.

In 2001 the British government created a Sustainable Development Commission to advise on policy in matters of sustainability. The Commissioner was Tim Jackson, Professor of Sustainable Development at the University of Surrey. His report *Prosperity without growth? The transition to a sustainable economy* was released in 2009 as an outcome of the Commission's work but it was written by Jackson as 'an individually authored think-piece'. It attracted lively interest in the UK and elsewhere and was quickly published as a paperback, now in its second edition. A review in the *Financial Times*, by an editor who had also served on the Commission, described *Prosperity* as a 'challenge to mainstream economic thinking that is both accessible and robustly argued'. There was no criticism of Jackson's 'beguiling vision' but as the reviewer saw it: 'The problem comes with reality' and the policy prescriptions suggested by Jackson 'take only the smallest of steps towards the post-growth society that he suggests we need'.[51] To be fair to Professor Jackson, he made no claim to have gone much beyond that, and he has readily admitted that 'there is as yet no fully fledged macroeconomics for a post-growth economy'.[52]

The way in which 'growth' is often used as if it were synonymous with 'progress' or 'improvement' is particularly annoying to some people. In Europe, an anti-consumerism movement marches under the banner of *décroissance*, a French word that could be translated as 'decrease' but which has been welcomed as 'degrowth' by English-speaking followers.[53] But although the new word highlights the opposition to growth it also invites questions about the end-point; economies can neither grow nor degrow indefinitely.

Advocates of degrowth seem not to have reached a consensus on a number of issues but if the general idea is to abandon growth as an economic objective, then for the time being it rests solidly, along with the steady state, at the bottom of the 'too hard' basket. Or perhaps some regard it as a 'too late' basket.

21 Food production

Agriculture did not replace hunting and gathering in the way that the internal combustion engine replaced the horse. Modern agriculture has not broken its continuity with the past. Crop species, like their wild ancestors, still need living soils of the right kind, nutrient cycling, appropriate temperature range, suitable water regime, pest control and pollination. Not so much a combustion engine, then, as just a different kind of horse.

It is in the sphere of agriculture that the interplay between ecology, economy and growth is brought most sharply into focus. According to FAO, in only three decades from now the world will need to be producing nearly 50 per cent more food, animal feed, and biofuel than it did in 2012.[1] This is to be done while attempting to avoid further damage to a planet that is losing an estimated 24 billion tonnes of soil each year, and has shown, during the past two decades, 'persistent declining trends in productivity' over 20 per cent of its surface. As for water, more than 1.7 billion people already live in river basins where water use exceeds the rate of natural replenishment, and at least 3,700 dams are planned or under construction, which will further reduce the remaining free-flowing rivers by more than a fifth.[2]

Needless to say, neither FAO nor any other authority on food security is suggesting that the world's people will all be adequately fed by mid-century if existing trends and practices continue. Nor has anyone discovered an easy solution to this gigantic problem. The green revolution was our answer to the food crisis of the 1960s but the green revolution has now run most of its course.

The present challenge is much bigger and there will need to be different regional mixes of technical, economic and behavioural responses to it.[3]

The question whether we are likely to be able to feed ourselves after a few more decades could have been asked ten thousand years ago. We have come a long way since small bands of people roamed around their own separate worlds but, looking back, we do not see a journey guided by collective wisdom and planning. It would be more true to say that we have been swept along by changes that simply accumulated. We now have unprecedented channels and opportunities for global cooperation but to a large degree the nations of the world are still being propelled along by rivalries, enmity, reactions and responses rather than any collective purpose.

The European Union has achieved an impressive degree of integration but it is still based on voluntary membership by nations whose leaders believe they are acting in the national interest. The phrase 'think globally, act locally', whatever its origins, has been used in an environmental sense since the 1960s, and as working principle with broad application it remains the best we have. Acting locally is usually the only option available to farmers but at least we can all be a lot better-informed in our global thinking than we were in the 1960s.

The current state of farming

Agriculture is the cause of an estimated 80 per cent of deforestation, and it releases about 24 per cent of greenhouse gas emissions measured as CO_2 equivalents and caused by human activity. That is because agriculture releases the largest share of methane and nitrous oxide which are much more potent than CO_2 as greenhouse gases.[4] Passions are also raised by the injustice of 'land grabbing' that results in the eviction of small farmers in favour of big commercial operations with their vast plantations of single crops. But it is all too easy to oversimplify the problems so it may be useful to devote a few paragraphs here to some basic facts and figures from acknowledged authorities who are most likely to have compiled them dispassionately.

More than 90 per cent of the estimated 570 million farms worldwide are family farms.[5] In 2014, FAO published an estimate that family farms occupied about 75 per cent of the world's agricultural land but a later analysis, while agreeing with the total number of farms and proportion of family farms, used additional data with a more conservative definition of family farms and concluded that they covered at least 53 per cent of agricultural land.[6]

Other than being in family hands, these farms around the world are so diverse that they may have little else in common. One tragic statistic that does link them, as revealed by numerous reports from nations rich and poor, is the high rate of suicide. Economically, farming is notoriously uncertain and it can be a socially isolating way of life.

In low to middle income countries the great majority of farms are smaller than two hectares but they employ most of the working rural population and provide much of the food for farming families. However, in some parts of the world, as a result of shared inheritance, these farms have been repeatedly fragmented to the point where many have become too small to be supportive unless family members have off-farm work. This situation was recently documented for North Africa and the Near East but it is much more widespread than that.[7]

In the USA, the median (midpoint) size of holding is around 500 hectares but the mean (average) size is about 100 hectares because of the large number of very small farms. Economically, nine out of ten farms are considered small because they have a gross cash farm income of less than $350,000. Despite their number, these farms accounted for only 24 per cent of the value of production in 2015. Large-scale family farms with at least a million dollars in gross cash farm income numbered fewer than three in a hundred US farms but contributed 42 percent of total production.[8]

For the European Union there are detailed census data for 2013 showing that more than ten million family farms manage 67 per cent of the agricultural land. These farms vary widely in size and earnings, both within as well as between countries. The average size of holding ranges from a little over one hectare in Malta to 72 hectares in the UK. Corporate farms amounted to only 2.7 per cent of farms but managed more than a quarter of the agricultural land—an indication of their much larger size. Corporate farms were most common in France, and in terms of output value, France is the biggest agricultural producer in the EU.

More than half of the farmers in the EU were over the age of 55 and most of those were farming the smaller holdings. These farms tend to be merged with other farms when the farmer retires, so the trend is towards fewer farms with a bigger average size.[9]

As to the income of farm workers, the EU statistics show that the highest incomes are earned in farms specialized for the production of pigs and poultry, while the lowest earnings are found among workers in the mainly small farms with mixed crops and livestock. In euros the average income in 2016 was less than € 10,000 in fourteen countries and it reached € 37,000 only in Spain and the Netherlands; a figure still well below average wages in the economy.[10] Overall, only 24 to 35 per cent of farms had a net farm income higher than their estimated opportunity costs—which is to say that most farms would do better to have their resources in a different enterprise if alternatives exist for them.[11]

Intensive farming

The Netherlands has excelled in developing and applying cutting-edge technology to the production of tomatoes, lettuce and other plants inside enclosures that are equipped like laboratories rather than greenhouses. Water and nutrients are precisely monitored and wastage is minimal.[12] This enclosed, intensive approach for crops is highly successful and generally admired but when similar production principles are applied to livestock there are completely different problems to be solved, quite apart from the abhorrence felt by people with a concern for animal welfare.

Pigs and poultry have long been produced by so-called factory-farming methods, and in the USA and elsewhere most beef cattle are fattened ('finished') in confined animal feeding operations (CAFOs) where they are fed on grain, mainly corn. Common to all these operations are the health hazards created by air pollution and the disposal of animal waste. As well as unpleasant smells and ammonia, the air carries organic particles from animal bedding, feed, skin, hair or feathers and it also includes potentially harmful bacteria and viruses.[13,14]

Worse still, farmers commonly administer low doses of antibiotics to intensively raised animals, both to reduce the chance of infections and to promote growth. The use of antibiotics for growth promotion was banned in the EU in 2006 but is still common elsewhere.[15] Exposure to low doses of antimicrobials selects for drug resistance among pathogens so airborne dust (as well as dung) may carry bacterial strains that are resistant to the same groups of antibiotics that are used to treat infections in humans. We'll return to this topic later in the chapter.

In traditional farming, animal dung from livestock is processed naturally in the fields. It becomes fertilizer so from an economic point of view that is an ecosystem service. But natural processes are soon overwhelmed if vast quantities of animal waste are concentrated in a small area. Sooner or later, depending on soils, slope, drainage, rainfall, duration of freezing and distances involved, excess nutrients, notably nitrogen and phosphorus, end up as nitrates and phosphates in groundwater, rivers and lakes. This nutrient pollution is ecologically disruptive and is a hazard to human health.[16] The same principles, of course, apply to human waste from towns and cities but the hazards are greater because water and toxins from many sources flow into sewers. Sewage sludge is generally subject to more rigorous testing and treatment before being applied to farmland as fertilizer.

In disposing of livestock waste, farmers are not likely to spend more money and time than is necessary to comply with local regulations, and they vary widely around the world. In some countries, including certain American States, the health and safety regulations are such that pig slurry from enormous vats ('lagoons') can be drifting about in the breeze from where it is being legally sprayed over fields.[17]

In the Netherlands it is to be expected that health and safety standards will be among the highest that exist. In that small country more than 12 million pigs, about 4 million cows and 100 million poultry birds are kept in intensive rearing facilities located among people living at a density of over 500 to the square kilometre. All manure must be processed in some way. Poultry manure is dried and incinerated; pig slurry is processed and the solid output is exported as fertilizer. But there are lawbreakers everywhere and manure fraud is a serious concern in the Netherlands. In a report to the Dutch parliament in 2016 it was confirmed that both nitrate and phosphate levels were too high, and that more efficient fighting of manure fraud would help in the effort to reach target levels.[18]

Fishing and aquaculture

Fraudulent documentation is also extensive in the fishing industry but the global trends are plain to see. Compared with the 1950s the world's fishing fleets have ten times the fishing power and yet the catch per unit effort has continued to decline. In 2014 there were estimated to be about 4.6 million fishing vessels in total, three quarters of which were Asian. With the same fishing power that was used in the 1950s the global catch would now be half what it was.[19]

In contrast, the growth of fish farming has been increasing in recent decades and in 2014, for the first time, aquaculture provided more fish for human consumption than did capture fishing in seas and inland waters. The outlook, based on present trends, is for aquaculture to continue growing strongly, with perhaps a slight increase in capture fisheries if some stocks are able to recover.[20]

Aquaculture is not a new idea; it has been part of rural life in China for thousands of years and has been practiced for centuries in parts of Europe. In the ancient Chinese tradition aquaculture is integrated with livestock farming so that each is dependent on the other. A classic system has ponds stocked with carp and ducks so that the water is fertilized with duck droppings and the growth of algae and zooplankton provides food for the fish. The ducks get much of their food from natural sources such as weeds and insect larvae so inputs are minimal. Alternatively, the ducks can be replaced with penned pigs, though the pigs gain no benefit from the pond and need to be fully fed.

This sort of integrated aquaculture has been well researched and documented in modern times, and international aid agencies have actively promoted it in many Asian and some African countries. The results in sub-Saharan Africa have been disappointing, partly because new projects aimed at raising carp to feed the family do not inspire much enthusiasm, and neither do carp offer much as a local cash crop.[21] In parts of Europe, since medieval times, farm ponds were created and maintained primarily for holding water, so they were available to be stocked with common carp and fertilized with manure from livestock.

In parts of Germany, Hungary and Poland carp are now farmed in very large ponds and fed with commercial, pelleted feed. Common carp can digest and obtain energy from carbohydrate so cereals can still be a major ingredient. The fish are sold locally and extra revenue is earned from recreational anglers. These open ponds (some would call them lakes), with natural vegetation and associated biodiversity, are a recognized ecological asset but as businesses, particularly in Poland, they show little profit. In one study it was found that German farms received more in state subsidies than they did from the sale of carp.[22]

Globally, aquaculture now involves well over 500 species of fish, molluscs (oysters, mussels), crustaceans (shrimps, prawns) and seaweeds in more than 200 countries that are registered to provide statistical data to FAO. Although data reporting falls short of requirements, FAO is able to say that fish culture in inland ponds accounted for 65 per cent of the increase in fish production for the period 2005–2014, and it is aquaculture's biggest contributor to global food security and nutrition. About 90 per cent of the total aquaculture production takes place in Asia and most of that is in China. Globally, the average per capita consumption of fish from all sources is estimated to be about 20 kg per year.[23]

Carp are not the only group of undemanding freshwater species. Catfish farming, in fresh and brackish water, has been brought to a high standard in Vietnam from where fish fillets are exported to Europe, the USA, Russia and many other parts of the world. A different species of catfish is farmed in the Southern United States, and to compete with cheaper Vietnamese imports the

American industry has become technically more sophisticated.

Like carp, catfish occupy a low trophic level in natural food webs (chapter 2) so their diet can include a proportion of cheap ingredients such as corn screenings and rice bran. But they cannot be fed entirely on cheap products, and the cost of feeding even these species can be more than half the variable production costs of a fish farm. As people everywhere spend more money buying fish from supermarkets they also come to appreciate greater diversity in their food, so fish and invertebrates from higher trophic levels are also being farmed in huge quantities, at higher cost, and with more pressure on the sources of commercial feed.[24]

China, the undisputed giant of both capture fishing and aquaculture, is the world's biggest exporter of fishery products. The marine fishing waters of China have been depleted for decades, so since 1985 China has been developing a 'distant-water fleet'. That fleet had grown to an estimated 1,900 vessels by 2010, and according to the *New York Times* it numbered nearly 2,600 vessels in 2017. The fleet operates in coastal waters over most parts of the world but is especially active off the West African coast where the Canary Current creates one of the planet's richest marine ecosystems. Chinese vessels are not the only ones to be scouring the oceans for valuable fish but the size of the Chinese fleet sets it apart, as do the reported billions of dollars spent in subsidizing the industry.[25,26]

In Chinese traditional waters, trawlers that would once have been catching the fish of choice are now hauling in so-called trash fish amounting to almost half the total catch. A five-month study by Greenpeace found the amount of trash fish (about 3 million tonnes a year) to be equivalent to the entire annual catch of Japan's fishing industry. Trash fish is just a convenient term for the collection of small fish and invertebrates that remains in the catch after selected fish have been taken out. Sampling of trash fish revealed that well over a third of the species were potentially edible commercial types—mainly juveniles.[27]

The bulk of China's trash fish is used as feed in aquaculture, either directly or after processing into fishmeal. Chinese aquaculture is extremely diverse and the animals being fed occupy a range of trophic levels. Filter feeders, such as mussels, can get enough food out of the water without being fed, and some others need only dietary supplements, but carnivorous types in the higher trophic levels are given trash fish directly in addition to compound feeds as supplements. Since the 1980s the output of filter feeders has declined as the production of carnivorous fish has grown.[28]

It must always be true to say that a kilogram of predatory fish, let's say sea bass, represents the conversion of a lot more energy than does a kilogram of seaweed or cabbage, regardless of how they were produced. But in the context of human food supplies does that matter? Not necessarily. If we were taking care to harvest seaweed and bass from the seas without depleting the stocks, that is to say, within their sustainable yields, then nature could keep growing them indefinitely while we grew the cabbage. But the bass have been over-exploited and are now being grown in cages moored offshore. They are being grown as fast as they can be sold while trawlers comb the oceans for fish with which to feed them.

As aquaculture continues to expand, the demand for trash fish increases and so does its cost. The response to this has been to replace trash fish with increasing proportions of plant proteins and oils in the pelleted feed. The proportions that give best results depend on the species of fish being farmed. All proteins are made up of amino acids and the ones from animals are more important in the diets of some species than in others, but overall, the proportion of plant proteins and oils in fish feed is substantial and increasing. Important sources of them are soybean, rapeseed (canola), sunflower seed and groundnuts (peanuts). These are important ingredients of many items on supermarket shelves, and in the feeds used for intensive pig and poultry farming. Ultimately, all the demand falls on agriculture.

This raises the question of how aquaculture compares with pig and poultry farms in terms of the efficiency of producing food. The best-known measure of animal production efficiency is the feed conversion ratio (FCR). This is the weight of feed given to an animal divided by its gain in weight, so the lower the ratio, the greater the weight that is gained from the feed. By this simple measure the efficiency of species being fed in aquaculture is much the same as that of poultry. And both are more efficient than pigs and cattle.

But the FCR is very limited in what it tells us. It does not take into account the nutritional value of the feed, the nutritional value of the product, or the proportion of the weight gain that is edible by humans. When all these measures are taken into account it becomes clear that the different species being fed in aquaculture are not equally efficient as food producers. In a study to compare a few aquaculture species with poultry, pigs and cattle, it was found, perhaps surprisingly, that the best aquatic performer among those studied was Atlantic salmon. Poultry (broiler chickens) were equally efficient, followed by pigs.[29]

Salmon, aquaculture and pollution

Farming Atlantic salmon requires a high level of technical sophistication as well as suitable coastlines, cool water (ideally 8°–14° C) and a daunting amount of start-up capital. Payback time on initial investment in recent years has been about nine years.[30] This is an industry dominated by big companies. Norway is the main producer of farmed salmon, followed by Chile, Scotland, Canada and Australia. With technical assistance from Chile, salmon farming has also been introduced to China and the

Koreas where cages are submerged at some depth in an attempt to maintain the necessary water temperature.[31]

All the salmonids (salmon and trout), both wild and farmed, make up only 4.4 percent of global seafood supply but salmon farming is in the public eye in western countries and has attracted most publicity in connection with health hazards and environmental impacts. In fact, the concerns that have been raised (apart from the synthetic colouring of salmon flesh[32]) apply to all operations where aquatic animals are being fed and kept at unnaturally high densities in one place for long periods.

Fish waste falls through cages and accumulates on the sea floor where its ecological impacts depend on water currents and the local ecology. Effluent from onshore tanks can cause similar problems. Parasites and diseases are able to build up very rapidly, sometimes wiping out captive stock and posing a threat to wild fish. Farmed fish eventually become genetically distinct from wild stock as a result of selection (deliberate or not) for desirable traits such as fast growth and tolerance of crowding. These fish are typically at a disadvantage in the wild and can lower the fitness of wild stocks if they escape and interbreed. Escapes will always be a possibility. In January, 2018, off the coast of New South Wales, Australia, an unexpectedly violent storm released 20,000 predatory yellowtail kingfish into a marine park. The fish, with a value of US$1.6 million, escaped from sea cages that were regarded as 'fortresses'.[33]

Sea cages, along with ropes, pipes and other equipment, must be treated with antifoulants to keep them free of barnacles and algae. Copper is usually the active ingredient, alone or in combination with other herbicides, and concentrations of these substances can build up around fish farms to the point where they are toxic to non-target species in nearby waters. However, the question that most concerns people who are shopping for fish is whether the products of aquaculture are safe to eat and serve to their children.

As with other livestock, many of the species in fish farms are treated routinely to prevent health problems. In tanks, doses of disinfectants, pesticides and medications, including antibiotics, can be calculated according to volume and added to the water, but in open cages the treatments must be given in or on the feed. This is far from precise because not all the feed is eaten, not all fish will eat the same amount, and fish that are most in need of medication are unlikely to have the best appetites. After being treated with antibiotics different fish will have different concentrations of the drugs in their tissues.

Antibiotics are effective mainly against bacteria and are used much more sparingly than they were in the past. Salmon farmers would like to announce that they have stopped using them altogether but they are still in use as a last resort. In Norway, where the salmon industry is regulated and monitored rigorously by the Norwegian Veterinary Institute, a total of 212 kg of antibacterials were used to treat farmed fish in 2016.[34]

As to whether antibacterials and other biocides in seafood are harmful to humans depends largely on what has been administered and how much time was then allowed to pass before the fish were killed and prepared for sale. That in turn will depend on regulations and their observance in producer countries, and the effectiveness of inspections in importing countries. Chinese scientists, in 2016, found traces of 15 different antibiotics in samples of common foods in Shanghai. Antibiotics were detected in 52 per cent of the aquatic products tested, compared to 35 percent in pork and 22 per cent in chicken.[35] On the other hand, the US Food and Drug Administration states plainly that 'aquacultured seafood is safe to eat and is considered a healthy food choice'.[36]

Of far greater concern to the medical and biosecurity experts is the growth in antibiotic resistance among those bacteria that cause human disease. Writing in 2013, an international team of scientists called this a 'tragedy of increasing magnitude that gravely affects human health'.[37]

I mentioned in the first chapter that bacteria have evolved many different kinds of metabolism and I could have said that their versatility goes far beyond that. Microbes can sense, communicate and interact using nothing more visible than chemistry. Clusters of genes (known in the technical jargon as mobile genetic elements) can be passed from one bacterium to another not only when cells divide to reproduce but by methods of horizontal gene transfer between unrelated cells. Some mechanisms make it unnecessary for the microbes even to touch each other.[38] What is becoming increasingly clear to specialists is that the transfers can be accelerated by pollutants, including drugs, heavy metals and disinfectants in the environment.

Genes that confer resistance to these compounds are passed between populations of microbes as an automatic response to the presence of the chemicals in the microbes' habitat. Furthermore, multiple resistances can be passed on the same cluster of genes, and one type of resistance mechanism can work against different chemical compounds. Resistance to an antibiotic could therefore arise in a population of microbes that had not itself been exposed to that particular antibiotic.[39]

We unwittingly provide every opportunity for antimicrobial resistance to arise wherever we bring together bacteria, residues of antimicrobial drugs, disinfectants, pesticides, heavy metals and a rich supply of nutrients. Aquaculture provides such conditions, as do sewage-treatment plants, pig farms, poultry farms and CAFOs everywhere.

It would be pleasing indeed if this overview supported a plan to feed the world with traditional methods. Idyllic scenes of small family farms with mixed crops and livestock, the fields studded with reed-girt ponds interspersed with shelterbelts of local trees is not just a pastoral dream, it is based on traditions that proved to be ecologically resilient and productive through centuries of economic, technological and social change. Where the traditions still work they are to be treasured but the challenge of feeding the world of today and tomorrow must be met by facing the harsh realities presented by the facts and figures above, and they do nothing to support pastoral dreams.

22 A synthesis

It is clear that turning-points in the human story have occurred at different rates and in different parts of the world with very little evidence of a steady progression towards what we now call modernity. Many of the main elements of modern life such as different occupations, organized religion, hierarchies of power, high density living, and credit with interest, came together in the Fertile Crescent but another few thousand years had passed before economic growth began to show much change on a per capita basis, and when that did happen it was the result of new initiatives in another region.

Economies grew in size and sophistication in response to the needs and initiatives of settled communities. By one means or another, energy and resources were converted into subsistence and wealth, and exploitation and distribution were mediated by economies and administrations of that place and time. It does seem that Karl Polanyi had a point when he claimed that economies before the nineteenth century were embedded in prevailing cultures. There is an obvious difference between modern economies and those of the past that were mediating human ecology and serving to maintain a communal way of life in a particular place. It can now be argued that economic growth has become the Great Cause that must be served by the people. Governments try to manipulate society in order to promote economic growth—spend more, save more, have more babies—it is hard to avoid that versatile metaphor in which the tail is wagging the dog.

It is not even true to say that economic growth always benefits everyone indirectly. Consider the financial sector. During the golden age of economics the trade in securities was conducted through stockbrokers and the average holding period for a share was seven years. Since then the size of the financial sector has grown phenomenally in major economies—a trend known, for want of a better word, as 'financialization'. Computer screens now provide all the information, and high frequency traders (HFTs) use computers with the latest algorithms to search for opportunities and get the best deals. Stocks may be held for less than a millisecond. John Kay, in *Other People's Money*, tells us that in Britain and the USA, about half the dealing in stocks and shares is conducted by HFTs and the annual volume of payments processed in Britain is about forty times the nation's income.[1]

A number of different developments brought about this transformation. Advances in technology are obvious among them but the growth of markets in derivatives has also been central to it. A derivative is a contract that derives its value from the price or value of some other security that underlies it and the derivatives themselves can be traded. In this way layers of financial claims are built upon underlying assets such that 'the value of outstanding derivative exposure far exceeds the value of all the assets in the world.'[2]

On May 6th, 2010, a trillion dollar 'flash-crash' in the USA was traumatic for the people who knew about it while it was happening. The market crashed but then recovered over a period of about 36 minutes—one of the most turbulent periods ever experienced by US financial markets.[3] Experts still disagree about the cause of it but Professor Kay reminds us that when computers are using the most sophisticated algorithms to conduct high frequency trading, no person fully understands what is happening. His own questions, on the other hand, are very easy to understand: 'What is it all for?' he asks. 'What is the purpose of this activity?' To be sure, the traders make, and lose, large sums of money between themselves but their activity 'contributes little, if anything, to the betterment of lives and the efficiency of business'.[4]

And yet, without any kind of economic growth the world's poor will have no means of lifting themselves out of poverty. And to the extent that economic survival, whether personal or national, remains dependent upon repaying debts with interest, there will be no obvious way of escaping the treadmill of growth even if poverty is overcome. So growth is celebrated wherever and whenever it occurs, regardless of the fact that some of its supposed benefits are obscure or questionable, and much of the environmental harm that it does is far-reaching, potentially disastrous, and indisputable.

Of course it is equally true that the technology emerging from modern economies has brought wondrous benefits, and it can be argued that social harm is linked to bad policy and personal choices rather than to GDP and productivity more generally. On balance, then, given that the growth imperative shows little sign of weakening, should we be optimists or pessimists about its further consequences?

Without much difficulty, you can find books and articles accurately describing the wonderful progress that has been made in improving the state of human life. People are living longer, infant mortality is a fraction of what it was, there is much less poverty, people are better educated, there is more democracy in the world, we have made amazing improvements in communications and transportation—the list could go on and on. But it is just as easy to find depressing accounts of increasing conflict, incidence of natural disasters, homelessness, suicide rates, debt and bankruptcy, stress and the disintegration of social fabric.

While these differing viewpoints may reflect some optimism and pessimism among the authors, it soon becomes apparent that what separates the two camps most clearly is the time-scale in which they are thinking. A survey conducted in 2015 revealed that 65 per cent of Britons and Americans thought that the world was getting worse. Only four and six per cent of Britons and Americans, respectively, thought the world was getting better. And from a selection of 17 countries, only in China did more people think the world was getting better rather than worse.[5] I think it safe to assume that all these respondents would have been thinking in terms of current trends, and there would have been a completely different outcome if they had been asked whether the world was better or worse than it was a century or more earlier.

There are other considerations that would separate positive from negative responses. If we think about the quality of life during recent decades, the trends have been mixed; good for some groups, not so good for others. But in responding to the survey question above, if we take 'the world' to mean the natural world, then it is hard to find any positive trends at all. They are all negative. Of course the quality of our lives and the state of the natural world are not separate issues but it will be easier in the following paragraphs to think about them under two separate headings.

Growth and human well-being

We know that neither GDP nor per capita GDP were initially intended to be measures of human well-being and yet, in truth, on some important measures over the long-term, improvements in human well-being do correlate quite strongly with per capita GDP. In 2014, the OECD published a major report in which ten different dimensions of well-being were examined for the period between 1820 and 2000.[6] Some dimensions, notably education (measured as literacy and years of education) and health (measured as life expectancy and height) were found to correlate well with GDP per capita. We must bear in mind, though, that correlations alone tell us nothing about causation, and analyses of the available data were statistically challenging.

For surveys in present time there is an obvious way to avoid statistical problems with different indicators of well-being: people can be asked to make their own subjective assessments and rate their satisfaction with life just as they decided on the state of the world in the survey mentioned above. In 2017 the fifth *World Happiness Report* was published. This report used data from Gallup World Polls based on roughly 3,000 respondents in each of more than 150 countries. In addition to objective measures of well-being, the polls provide respondents' own evaluations of their lives on a ten-point scale where 0 represents the worst possible life and 10 means life at its best.[7]

On this scale of subjective well-being, the ten happiest countries are seen to be separated from the ten unhappiest ones by four points; that is to say the average life evaluations in the top ten countries are more than twice as high as in the bottom ten. This is not a difference that can be explained by per capita GDP or average incomes; it is based on a combination of economic, social and emotional variables and the economic ones do not explain the others. China demonstrated this particularly well. Between 1990 and 2015 there was a five-fold increase in GDP but subjective well-being declined until 2005, though it has improved substantially since then.

Six key variables account for three quarters of the differences among countries. They are GDP per capita, healthy life expectancy, social support (having someone to depend upon in times of need), trust (in government and business), freedom to make life decisions, and generosity. Having someone to depend upon appears to be especially important. Within a nation, if the proportion of people who have that support is increased by ten per cent, average life evaluations are increased by the equivalent of a doubling of GDP per capita.[8]

The top ten countries listed in 2017 were the same ones as in the previous year, and the top four (Norway, Denmark, Iceland, Switzerland) were still the top four, though not in exactly the same order.[9] All ten nations rank highly on the six variables listed above and none is among the world's biggest economies as measured by GDP in international dollars, although Canada and Australia do make it into the top twenty. Nor do any of the ten happiest countries have large populations; quite the opposite, all but two of them (Canada and Australia) have fewer than ten million people.

In summary, hundreds of thousands of people around the world have revealed that, while they appreciate having more money to spend, their overall satisfaction with life depends on a combination of economic, social and emotional variables, and simply having more money does not compensate for shortfalls in other areas. I doubt that many people will be surprised to hear that the recipe for sustained happiness comprises a number of ingredients, not just more and more of one of them. The track record for recent decades in the USA shows this clearly. In constant dollars, income per person has increased by roughly 250 per cent since 1960, but measured happiness has not risen and was recently found to be falling as per capita GDP was rising.[10]

In chapter 8 I mentioned Karl Polanyi's assertion that individuals covet expensive possessions only to the extent that they safeguard a person's social standing. If we take that as a sweeping statement to mean that we are obsessed by status symbols and a need to keep up with the neighbours, then it would surely be easy to refute. But think of it in terms of fair play and equality, and it could be that, once again, Polanyi had a valid point. It explains the so-called Easterlin effect—shown by those countries that have become richer and better educated without showing any increase in average happiness. The evidence suggests that the important thing is our wealth and education relative to that of others with whom we compare ourselves, rather than the actual measures in dollars and schooling. If advancement is unequal then subjective well-being is reduced for those who are left behind. If everybody moves upwards together then subjective well-being is not necessarily raised by the progress that is made.[11]

More detailed studies are needed to tease out the influence of different components of broad variables such as 'healthy life expectancy'. Major surveys have been conducted in four countries: USA, Australia, Britain (two surveys) and Indonesia. In all three western countries it was found that common mental illness (diagnosed anxiety and depressive disorders) was more important for happiness than income, employment or physical illness, and that treating depression and anxiety is by far the least expensive way of reducing misery. In Indonesia, income emerged as being more important than mental health, though that, too, was important.[12] Worldwide, according to the World Health Organization, depressive disorders are the single largest contributor to non-fatal health loss.[13]

Whether mental health decline is connected with increasingly stressful living is difficult to say; hard evidence is less easy to find than are opinions in the popular media, but there are physical health problems that can certainly be attributed to modern living. There was a time when obesity was associated mainly with poor diets and overeating by individuals in wealthy consumer societies but the incidence of obesity has tripled since 1975 and it is now also a problem in low to middle income countries. This is especially so in urban centres where diets are based on selections from supermarkets rather than on the produce of family smallholdings. Nearly half the children who were overweight or obese in 2016 lived in Asia, and according to WHO, being overweight or obese is now linked to more deaths worldwide than is the opposite condition of being underweight.[14]

It is easy enough for anyone to see both good and bad in the social consequences of economic growth and I shall not devote more space to it.

Growth and the environment

The environmental impact of economic growth has not been a mix of good and bad; it has been consistently detrimental, although countries that have grown richer can more easily afford to repair some of the damage. The story of a curious little animal called the axolotl (*Ambystoma mexicanum*) exemplifies a number of important realities of ecology in the modern world.

The tale of the axolotl

Axolotls are amphibians that once lived out their simple lives in Lake Xochimilco, part of the lake system of the Valley of Mexico. They continued to thrive in the canals and wetlands that the Aztecs engineered after they founded Mexico City on an island of the central lake. The axolotls were collected and eaten by local communities for centuries but they survived at densities of thousands to the square kilometre. Nor were axolotls in any real trouble after the Aztecs were conquered by the Spanish and allied forces in the sixteenth century, though the wetlands did begin to deteriorate ecologically. But Mexico City more than trebled in size between 1950 and 1975, and the urbanization put an unprecedented strain on the supply of fresh water. That and other changes have proved too much for axolotls.

Today, all that remains of their habitat are the polluted canals of a southern district of Mexico City, where introduced European carp and African tilapia are known to prey on axolotl eggs and young. In 2006 the axolotl was listed as critically endangered by the IUCN.[15,16]

In its broad outlines this could have have been the story of any number of obscure creatures whose habitat has been destroyed by human activities, and which now survive as remnant populations on their way to extinction. But in the case of the axolotl we had the chance to discover its particular importance to us. Zoologically, it belongs to a group of salamander species and subspecies but is unique among them in that it lives and breeds in the aquatic larval form. Only rarely does it metamorphose into a terrestrial salamander. Moreover, it has outstanding powers of regeneration and can repeatedly regrow severed limbs and some other structures without so much as a scar to show for it. Nor are there problems of rejection when axolotls receive organ transplants from other individuals.

Not surprisingly, axolotls are valued for research into tissue regeneration and certain other fields of medical biology and are common in laboratories around the world. They are also quite popular among aquarium enthusiasts so there is no shortage of captive specimens. But captive axolotls are seriously inbred and many carry the genes of another species with which they were cross-bred at some point. Captive populations, therefore, are not a suitable source of specimens for reintroduction to the wild even if healthy natural habitats can be found for them. This, too, is a familiar problem of conservation: captive populations can save a species from extinction but deliberate, and often expensive, management is needed if the aim is to reintroduce the species to the wild. And the wild must be ready to receive them for there is no point in returning animals and plants to the wild in order to confront whatever it was that threatened their survival in the first place. And that brings us to the subject of protected areas (PAs), the cornerstone of conservation strategy.

Protected areas

The Biodiversity Convention came into effect in 1993 and the tenth meeting of the parties was held in 2010 in Aichi Prefecture, Japan. It was at that meeting that a Strategic Plan for Biodiversity was adopted in an attempt to halt the losses and ensure that resilient ecosystems are secured and will continue to provide essential services. Some PAs contain the ancestral types of the crops that we now depend upon, and are therefore sources of the original, sometimes precious, genetic diversity of the species.

Among the twenty Aichi targets that were adopted for the 2011–2020 period, is that of securing systems of well-protected areas that are integrated into the wider landscapes and seascapes (target number 11). More specifically, the systems should be ecologically representative, well-connected, and include at least 17 per cent of land and inland waters, and 10 per cent of coastal and marine areas.[17]

So far, just under 15 per cent of land and inland waters, and just over 10 per cent of coastal and marine areas are said to be protected. At first glance, these figures look quite encouraging, but we all know how deceptive first impressions can be. The location of protected areas, as well as their extent, is crucially important because as I mentioned in chapter 15 the world's biodiversity is not at all evenly distributed. In fact, less than 20 per cent of key biodiversity areas are completely protected, and only ten per cent of terrestrial ecoregions have more than half their area protected.[18]

Areas that are least affected by competing forms of land use, such as Greenland and remote parts of the oceans, are among the easiest to bring under some form of protection but the richest areas of biodiversity tend to be in warm, fertile regions and coastal areas where land-use conflicts are most intense. Not only are PAs most difficult to create in such circumstances, they can be notoriously difficult and expensive to manage. It is often pointed out that indigenous peoples and local communities have been managing biological resources for millennia, and not only for subsistence but also for cultural and spiritual purposes, so people from those groups should be the obvious custodians of modern PAs.

It is fortunate indeed when local knowledge and commitment can be part of PA planning and management, so much so that the participation of indigenous and local people is targeted under the Strategic Plan (target 18), as is the flow of benefits to them from PAs. But the fact that a particular group of people have supported themselves for millennia in their natural environment does not, in itself, solve any problems caused by more recent developments such as agriculture, the use of chain-saws, firearms, wire snares, outboard motors and an expanding trade in bushmeat and forest produce. Nor does it necessarily reduce the external threats from mining, industrial agriculture, industrial fishing or urban sprawl. For each PA there will be a different mix of economic, ecological and social realities together with local laws and traditions of land use and tenure. The importance of local participation cannot be overstated, but the opportunities and constraints are endlessly variable.

Protected areas may have no tourist potential and few benefits that can be shared directly with local people. Others may provide invaluable ecosystem services to communities at a distance; it is reported, for example, that one in three of the world's 100 largest cities receives drinking water from PAs.[19] In other cases, wild nature may be an enormously valuable tourist attraction as in some of Africa's famous conservation areas or India's tiger reserves. Yet even relatively well-funded PAs such as those can fail to reconcile conservation with the growing and changing demands of local communities.

Integrated conservation and human development projects have had mixed results and have proved difficult to evaluate.[20] In theory, buffer-zones adjacent to PAs can allow access to some wild resources while prohibiting destructive categories of land use, but unless the usage is monitored, and the rules strictly enforced, the buffer zones end up being indistinguishable from the agricultural land beyond. Protected areas can come to resemble natural oases in a landscape of cultivation or overgrazed pasture, and that makes them irresistibly attractive to land-hungry subsistence farmers.

When buffer zones are gone, foraging for bush materials, trapping and snaring of animals, and even clearing and settling begin to press directly upon more protected zones. The core areas of some well-known parks and reserves have proved so difficult to protect that large mammal densities are well below carrying capacity in many African and Asian National Parks.[21,22] India's famous Kaziranga National Park is a carefully-managed World Heritage site in Assam, but there were violent confrontations in 2016 when authorities used elephants to destroy the huts of settlers far inside the park boundaries.[23]

Among the thousands of PAs that are less well-known, many are protected on paper only and they lack the staff, funding and political support to be anything more. The decline in biodiversity would be much worse if we did not have protected areas but far too many of them are failing to prevent serious losses and it is obvious that their number and combined area are no indication of their effectiveness. Effectiveness needs to be separately assessed and it can be done only if the resources are available.[24,25]

I have often wondered how people of the distant future will look back at our present time. Will they be impressed by how much was achieved during a couple of centuries or will they shake their heads uncomprehendingly at our blunders? What sort of turning-point are we living through? We are uncertain about our food security, we have destroyed much of the fabric of the natural world that supports us, we have chained ourselves to engines of economic growth that only partially serve our needs and we have upset the natural controls of the climate. The destructive trends are continuing and even accelerating as our numbers increase.

The journey continues

Despite the uncertainties and the eternal need for more research, we do have enough information to guide us in the general direction of sustainability, even if we are doomed to keep growing the economy in some ways. And many of the measures needed to reduce one form of pressure will often have wider benefits for the natural environment, though that is not always the case as I shall try to make clear under the headings that follow.

Food security

The aim is to have more food without converting much more land so a reduction in losses and wastage would help enormously. To get seeds or fruit we must grow a whole plant, and crops can be lost because of a single storm or flooding rain, but avoidable losses also occur in harvesting, transporting and storing of food. Wastage occurs when unsold food is dumped and when food is bought but not eaten. Best estimates suggest that 31 per cent wet mass of crops is lost between harvest and consumption.[26]

It can also be argued that over-consumption is a form of waste if significantly more calories are used up than are needed to maintain good health. A recent analysis suggested that the waste of over-consumption is at least as great as that from food discarded by consumers. And the same study found that only about 12 per cent of the calories in crops fed to livestock are converted to meat in the human diet. This agrees quite well with the ten per cent rule of thumb that I outlined in chapter 2.

Food from animals provides about 25 per cent of the protein

consumed by people globally and it is not easy for everyone to be a healthy vegetarian. We are omnivorous animals, and meat is not only high quality protein but a source of micronutrients such as vitamin A, vitamin B-12, riboflavin, calcium, iron and zinc. Vegetarians in cities can meet their dietary needs but billions of others would find it difficult to survive on locally available plant products. Beef is a particularly extravagant food in terms of land use although it is not as wasteful as critics claim when they assume that cereals could have been grown in place of cattle. Much grazing land is not at all suitable for growing cereals or other crops. According to FAO, of the total grassland area used by cattle and buffaloes, only about 27 per cent could be used for crops.[27]

Recognizing that the world is in a 'post green-revolution' era, FAO and major aid donors have been urging farmers to combine traditional farming with other practices that will conserve water and enhance the natural productivity of soils. This might seem to be going backwards but if traditional practices such as crop rotation and mulching are guided by the findings of modern science then farming is on the path of 'agroecology' or 'conservation agriculture'. The two terms mean more or less the same thing, though some theorists quibble about definitions.

Basically, the approach comprises three principles of management. Soil disturbance is kept to a minimum by reducing or eliminating ploughing and other forms of tillage; soil loss is prevented by growing cover crops or mulching, and different crops are rotated to prevent the rapid build-up of pests and to spread nutrient cycling through different root zones. This can involve growing a variety of annual and perennial plants in associations or crop rotations that might include trees, shrubs and pastures. Plants that fix nitrogen in the soil (mainly legumes) have an important role and can be incorporated in the scheme either as scattered trees and shrubs or as part of a crop rotation.[28]

All this is music to the ears of ecologists but how well does it fit with those harsh realities that I outlined in the previous chapter? Sadly, not nearly well enough. The diversity of farms and farming around the world is such that the best prescription for conservation agriculture will not be the same from one farm to another. Among the big, highly mechanized farms of Australia, Canada, the USA, South America and parts of Europe, the value of no tillage was recognized decades ago as a way of saving time and tractor fuel. It proved more economical to clear weeds with a herbicide before sowing a crop. Glyphosate has been the herbicide of choice but its exclusive use has selected for resistance among weeds. Glyphosate resistance is now such a problem that some farmers in the USA have had to return to ploughing. It is reported that 38 species of troublesome weeds are now resistant to glyphosate in 37 countries.[29]

In contrast, on the poor red soils of Africa where it is difficult to grow any crops without applying fertilizer, conservation agriculture on a small farm without mechanization means that more time is spent in hand-weeding, and the burden commonly falls on women. Mulch from crop residues, (if not fed to the cow and donkey) becomes fast-food for termites and there may be a shortage of manure from livestock. There will certainly be no land to spare for trying out different regimes to see what works best in the longer term. Where conservation agriculture has produced good results it can have taken four to six years to find success.[30]

A meta-analysis, published in 2015, was carried out using results from 610 studies across 48 crops and 63 countries. It was found that no tillage without the other two components of conservation agriculture produced different outcomes but on average it reduced yields.[31] The take-away message from another international team of scientists is that if agriculture in all its forms is to be intensified sustainably, then flexibility and a 'place-based' approach is needed, rather than dogma and prescription.[32]

The Chinese evidently reached similar conclusions some time ago. In the decade between 2005 and 2015 a nationwide project was carried out involving 21 million smallholders and over 13,000 field studies on corn, wheat and rice crops. The findings were relayed to small farmers who increased crop yields by an average of 11 percent while using about 15 per cent less nitrogen as fertilizer. Nitrogen is a source of the greenhouse gas, nitrous oxide, and China's use of nitrogen per hectare is about four times the global average.[33] It was a project on a vast scale and the spread of the research network was key to its success. Tens of thousands of local officials and even greater numbers of representatives from agribusiness were in some way involved in the work of about 1,200 scientists. Small farmers were kept informed through outreach programmes and thousands of workshops. Other countries would find the scale impossible to emulate but the project offers some useful lessons.

Biotechnology may hold the greatest promise for improving food security, and spending on research and development in agriculture has been rising since the turn of the century after a slump during the 1990s.[34] But the controversy about genetically modified organisms (GMOs) has polarized opinions about the whole subject. Nevertheless, In 2014, GM crops were being grown in 28 countries by eighteen million farmers.[35] Genetic modification, or genetic engineering, involves transferring genes from one species to another where the species are not genetically close enough for conventional cross-breeding. The risks and benefits need to be thoroughly investigated case by case. To regard all GMOs as good or bad is a little like thinking that all medicines are good or bad instead of weighing the need for each medicine against the risk of unwanted side-effects.

Biodiversity

Nothing can replace a comprehensive global network of adequately protected areas but that is not to say that nothing else is worth doing. Local community groups and individual landowners undoubtedly make a valuable and commendable contribution to protecting local fauna and flora. Properly-managed sport-hunting can also be sound conservation if it prevents natural ecosystems from being converted, and provided that offtakes and population responses of trophy animals are carefully monitored and controlled. I can declare no conflict of interest here, never having shot an animal for sport in my life.

It is sometimes inevitable that wildlife management will be in conflict with animal welfare because the aim of conservation is to prevent losses of species and habitats whereas animal welfare is focused on the well-being of animals as individuals. Opinions are similarly divided on the role of zoos. In some parts of the world zoos are still nothing more than miserable prisons for animals but modern zoos uphold higher standards. They do what they can to prevent extinctions of charismatic species and they use their opportunities for research and conservation education. But zoos are usually dependent on gate-money for their survival so directors would be foolish to neglect a zoo's recreational appeal.

Modern zoos are generally linked through regional federations, which in turn are members of a global body—the World Association of Zoos and Aquariums (WAZA). Endangered species can be managed in breeding programmes for which stud books are maintained and animals are selected for breeding with the aim of preserving the species' genetic diversity. Botanic gardens serve similar purposes for plants and are also linked though a global network called Botanic Gardens Conservation International. If the world's natural wealth is to be brought through this century without further substantial losses, then conservationists surely need access to every tool in the box.

Commercial energy

In round figures, fossil fuels make up about 80 per cent of the world's use of commercial energy. As a proportion of the total, a little less than one fifth is used for generating electricity and the rest is used for transport, heating and other services.[36] The challenge for the world is to make the transition from fossil fuels to renewable energy technologies in order to reduce greenhouse gas (GHG) emissions.

It is technically possible to capture and store carbon dioxide from burning coal, natural gas or biomass, but in practice it has so far proved to be too expensive and problematic.[37] Accordingly, renewables are the world's fastest growing energy source but total energy use is also growing strongly and it is predicted that fossil fuels will still account for three quarters of the total in 2040.[38] We often see figures published for the growing proportion of renewables being used in power stations and should bear in mind that the proportion of primary energy used to generate electricity will also be growing as the developing world builds more power stations.

Renewable energy technologies are limited in where and when they will work and they are not without environmental impacts of their own. There are six of them in addition to storage technologies: wind, solar, hydroelectric, biomass, geothermal and tidal.[39] Their efficiencies can be measured in different ways. Energy payback time (EPBT) is the time it takes to recover the energy invested in a system during its lifetime, taking into account the energy used in construction, maintenance and decommissioning. Working all that out is a task known in the industry as a life-cycle analysis.

A more direct and frequently published comparison of energy sources is based on the energy going into, and coming out of, a system without considering payback times. In other words, how much usable energy is gained from a source, relative to the amount of usable energy that is spent in producing it, regardless of the time factor. This ratio is usually expressed as the 'energy returned on energy invested' (EROEI), or 'energy return on investment' (EROI) for short. A fuel with an EROI ratio of 1:1 would mean that energy obtained would be equal to the energy used in releasing it so the entire exercise would be a waste of time. A fuel with an EROI of 2:1 still only delivers 50 per cent of its energy as a net gain so energy sources with higher values are essential for modern societies.

With fossil fuels the usual trend is for EROIs to fall as the most accessible deposits are used up and extraction gets progressively more expensive. In the case of oil, as the cost increased it became economically viable to extract and upgrade the products of oil shale and tar sands by processes that are environmentally more destructive than is oil drilling or coal mining.

There are wide variations in published EROI values for renewable energy sources, but recent studies have produced values of about 4 for solar and biomass compared with 16 for wind, 30 for coal and 49 for hydroelectric. Nuclear power has been rated as high as 75 but much lower values are reported when the cost of decommissioning is included. Similarly, the values for solar and wind, as intermittent sources, are lowered further if the cost of storing energy is taken into account.[40,41]

Manufacturing costs for solar and wind power have fallen sharply in the last decade and should fall further as technology improves but economies of scale will be limited for renewables where siting is important. Large scale expansion will inevitably mean that successively less favourable sites are used as the best ones are taken. In particular, hydroelectric power appears to be a very attractive proposition but only the best sites produce such high values and the sites of easy access for large-scale projects are already in use in most developed countries.[42]

The minimum EROI to support a nation or society (EROIsoc)

depends on the society's energy needs at a particular time. In calculating the EROIsoc for a nation the 'energy returned' is the average amount of energy that can be bought with a dollar in that nation. Energy invested is the primary energy needed to produce a dollar's worth of economic production—the energy intensity of the country. In low income countries the EROI may be lower than 4 whereas in upper middle-income countries it can be well over 20. A ratio of about 7:1 is widely thought to be the economic break-even point in sustaining a modern socio-economic system.[43,44]

There is no quick way of making the transition from fossil fuels to sources with much lower EROI values. Most experts think in terms of working towards a mix of energy sources but nations vary in their existing energy status and have different opportunities for change. Australia, for example, is particularly well-placed to expand solar and wind power, and costs for generation are said to be relatively low. A recent study concluded that if supportive government policies were in place, 96 per cent of all energy use in 2050 could be coming from a mix of renewables.[45]

The European Union is on track to have renewables making up 20 per cent of its 'gross final energy consumption' (including electricity) by 2020. The main contributor to the renewables at present is wood and other solid biomass (including waste) but that can result in more land being being converted for tree plantations and an increased threat to natural forest. The energy is renewable only if wood is used within sustainable yields. Carbon is released in the burning, as with coal, but in time it is absorbed again as replacement trees are grown so productive burning is simply inserted into the natural cycle. Well, that's the theory but Europe's dependence on biomass has some very outspoken critics.[46]

Whatever the mix of energy sources, environmental impacts will be unavoidable so the aim must be to keep the impacts to a minimum. Dams that block and divert rivers for hydroelectric power can be devastating to local ecosystems and in some cases to human communities. The vast Three Gorges Dam in China is reported to have displaced nearly 1.2 million people. Using the weight of falling water to drive turbines does not produce greenhouse gases, though large areas of flooded and rotting vegetation can release methane for a long time, as well as killing the fish. In South America 80 per cent of the potential for hydroelectric power is located in rain forest.[47]

Wind turbines are known to kill birds and bats in quite large numbers but not, as far as is known, to the extent of affecting their populations. The greater potential for ecological damage lies in the construction and siting of the giant turbines. The tower for a turbine may be set on a thousand tonnes of concrete, or some combination of less concrete and more heavy steel anchor rods. As in all renewable technologies, fossil fuels are used in the manufacturing and decommissioning stages but in the case of wind turbines there is also a high risk of ecological damage if the construction sites are not carefully selected.

The turbines in a windfarm are separated by distances of 5 to 10 rotor diameters and the area must be kept free of trees and allow for easy access, though livestock can be grazed. And as well as the excavations for concrete bases and cable trenches there are serious roadworks to be completed before the massive components can be delivered to the site. On fragile soils such as the peaty uplands of Britain the environmental consequences of such disturbance can be surprisingly far-reaching.[48]

By comparison with wind turbines, solar (PV) panels are the essence of convenience and if they are fitted on existing buildings they do not even add their own footprint. But how do PV panels compare with wind power in reducing greenhouse gas emissions, given that they both use fossil fuels during manufacture, setup and decommissioning? In one recent and revealing study it was found that average life-cycle greenhouse emissions for wind energy were 34.1 grams of CO_2 equivalent for each kilowatt-hour of electricity supplied. The comparable figure for PV panels was 49.9. But there was a wide disparity around these averages as they are influenced by at least eight variables including technology, sizing, longevity and calculation methods.[49] As a point of interest, the authors of this study ask which will have the lower emissions of greenhouse gas: wind and solar with storage and a lesser need of fossil fuel, or wind and solar without storage and a greater need of fossil fuel? They can offer no answer as the question has scarcely been addressed in the research literature. Clearly, there is a lot more work to be done.

But wind and solar energy can be harnessed all over the world (in contrast to the limited opportunities for geothermal and tidal power) and will be essential in the struggle to slow the change of climate. Much more questionable is the benefit of using biofuels. The main biofuels are ethanol (ethyl alcohol), which is mainly distilled from corn; and biodiesel, which can be made from plant and animal oils but is produced mostly from soybeans, canola (rapeseed) and palm oil. Biofuels, therefore, are in direct competition with food production and the supply of feed for livestock and aquaculture.

At first, biofuels seemed to be such a good idea that governments began to mandate their use as a regulated proportion of transport fuels. By using living biomass rather than the fossilized kind it can be assumed that CO_2 released in their manufacture and burning will be soaked up again in growing the future crops. If only it were that simple! In 2010 the US Environmental Protection Agency calculated that corn-grain ethanol would have life-cycle GHG emissions higher than those of gasoline unless the biorefinery used biomass as a heat source, but it was predicted that there would be a 21 per cent advantage to ethanol by 2022. As for whether biodiesel made from oilseeds reduces

GHG emissions, the US National Academy of Sciences declared in 2011 that there was no scientific consensus on the question.[50]

Let us imagine that technology works new wonders in the coming decade and biofuels are shown to reduce GHG emissions by 20 or even 30 per cent. Will it be a worthy trade-off against the extra land needed to grow crops for biofuels? After all, the need for human food and livestock feed will also be growing. Writing in 2016, John DeCicco, research professor at the University of Michigan, left us in no doubt about his own conclusions. The promotion of biofuels has been a blunder and as a cure for climate change, biofuels are 'worse than the disease', he wrote.[51] Authorities in the US, aware that clearing more natural ecosystems for bioenergy crops would release still more greenhouse gases, amended the Renewable Fuel Standard in 2007 to the effect that crops for biofuel would not comply with the Standard if the land for growing them had been cleared after 19 December that year.[52]

The European Union, in January 2018, announced its decision to cap all crop-based biofuels at 2017 levels, and to cease using palm oil as a transport fuel in 2021 because of its role as a driver of deforestation. Not surprisingly, major producers of palm oil reacted angrily to this decision.[53] Regrettably, in the exaltation of biofuels, a lot of damage has already been done.

Growth and demand

There are a few variations of the old joke about teaching a parrot to say 'supply and demand' so that it will be able to answer every question in economics. The joke makes a good point because the law of supply and demand is truly fundamental in economic theory. But to my mind the joke also makes the point that too much economic dogma is repeated, parrotlike, in the mantras of politicians, economists and the captains of industry.

We must keep growing the economy, they tell us, so that more jobs can be created for the greater number of people who will need them, and we must keep increasing the number of people because they are consumers who will expand our markets and grow the GDP so that more jobs can be created. More land must obviously be converted in order to feed, accommodate and generally cater for the growth in consumers, who will need more electricity, schools, hospitals, roads, infrastructure—and so *ad infinitum*. In the urge to keep the GDP ratcheting upwards there is no end-point in mind, no withdrawal strategy, and no patience with those who question the sense of it all. 'You can't stand in the way of progress' is one of the supporting mantras. As a goal for human progress, the perpetual growth of material things is not realistic but it clearly serves the interests of those who make money from paving its way for as long as they can keep it going.

Most of my generation, myself included, grew up believing the dogma because we had no counter arguments. We could not demonstrate that growth, in itself, does not make people happier, or that some of the happiest and economically most successful nations are among those with small, relatively stable populations, or that some economies could be growing while their populations were slowly declining. We had not even reached the stage where decoupling economic growth from resource use and population growth had been considered as a possibility. Historically, economies had always grown along with populations; that was the received wisdom.

And the received wisdom is remarkably resistant to change because it still serves the interests of exceptionally influential people, including media moguls. Take note and you will surely see more letters to the editor than news stories linking population growth with environmental ills. But in all fairness, as Robert Engelman observed in the *Scientific American*, a topic that includes sex and contraception, immigration, abortion and choice of family size is not the easiest to discuss objectively. 'What in the population mix is *not* a hot button?' he asked.[54]

And yet if we do face the facts objectively we see that the UN medium variant projection of 9.8 billion people by mid century is not a given; it is a projection based on present trends and policies. If fertility levels are consistently 0.5 babies below the assumption used for the medium variant there would be 8.8 billion people at mid century—fewer by a billion people.[55] We also know, from solid evidence, what can be achieved by family planning programmes when they have properly-trained personnel and financial support. There are 214 million women in the world with an unmet need for contraception.[56]

In Europe and some other economically advanced countries we have seen how a majority of couples have chosen to have much smaller families, despite government policies aimed at boosting the birth rate and preventing any decline, and we have good reason to believe that economies can adapt to slowly-falling numbers.[57]But is it likely that the world could adapt peacefully to the kind of growth that was reached during the golden age of economics, when the population was less than half its present size? More to the point, do we really want to find out? Abrupt reductions of numbers by famine, disease, disasters and strife are inevitably major tragedies in human terms, if not economic ones.

A turnaround in world population growth, even well before the end of the century, is obviously not a quick fix but it is a goal that will make short-term measures more effective and worth pursuing. Otherwise, we are applying palliatives in the hope of mitigating problems that we are continuing to multiply, and this is proving to be an ineffective response to the changing climate and the known ecological trends. On the global scale, it is nothing new for humans to be following or suffering change rather than directing it, but never before in the history of our species have so many people been crowded in harm's way.

Notes and References

Introduction

1. Robert J. Shiller, 'Is economics a science?', posted on *Project Syndicate*, 6 Nov. 2013.

Chapter 1 Primal growth and the single currency

1. John W. Valley et al., 'Hadean age for a post-magma-ocean zircon confirmed by atom-probe tomography', *Nature Geoscience*, vol. 7, 2014, pp. 219–23, doi:10.1038/ngeo2075
2. Until the 1970s these single cells were all called bacteria. They are now classed as either bacteria or archaea on the basis of their biochemistry. For convenience, I use 'microbes' where the meaning applies to both groups.
3. US Dept of Health & Human Services, National Institutes of Health, 'NIH Human Microbiome Project defines normal bacterial makeup of the body', news release of June 3, 2012.
4. J. Thomas Beatty et al., 'An obligately photosynthetic bacterial anaerobe from a deep-sea hydrothermal vent', *Proceedings of the National Academy of Sciences*, vol. 102, no. 26, 2005, pp. 9306–10, doi:10.1073/pnas.0503674102
5. Heinrich D. Holland, 'The oxygenation of the atmosphere and oceans', *Philosophical Transactions of the Royal Society B*, vol. 361, no. 1470, 2006, pp. 903–15, doi:10.1098/rstb.2006.1838
6. Nick Lane, *Power, Sex, Suicide: Mitochondria and the Meaning of Life*, Oxford University Press, Oxford, 2005, p. 132.
7. William Martin, and M. Müller, 'The hydrogen hypothesis for the first eukaryote', *Nature*, vol. 392, issue 6671, 1998, pp 37–41.
8. W.F. Martin, S. Garg and V. Zimorski,. 'Endosymbiotic theories for eukaryote origin', *Philosophical Transactions of the Royal Society B*, vol. 370, no. 1678, 2015, doi:10.1098/rstb.2014.0330
9. Nick Lane, 'Energetics and genetics across the prokaryote-eukaryote divide', *Biology Direct*, vol. 6, no.35, 2011, doi:10.1186/1745-6150-6-35
10. Nick Lane and W. Martin, 'The energetics of genome complexity', *Nature* , vol. 467, issue 7318, 2010, pp. 929–34, doi:10.1038/nature09486

11. Nick Lane's book (note 6) is outstanding as an authoritative yet very readable account of the importance of mitochondria in our lives.
12. Phil Jardine, 'Patterns in palaeontology: the Paleocene–Eocene thermal maximum', *Palaeontology Online*, vol. 1, article 5, 2011.
13. Kevin E. Langergraber, 'Generation times in wild chimpanzees and gorillas suggest earlier divergence times in great ape and human evolution', *Proceedings of the National Academy of Sciences*, vol. 109, no. 39, pp. 15716–21, doi:10.1073/pnas.1211740109
14. Arthur D. Chapman, 'Numbers of living species in Australia and the World', Report for the Australian Biological Resources Study, 2nd edition, 2010, Australian Government, Canberra.

Chapter 2 Thinking in terms of energy

1. Michael D. Sockol, 'Chimpanzee locomotor energetics and the origin of human bipedalism', *Proceedings of the National Academy of Sciences*, vol. 104, no. 30, 2007, pp. 12265–9, doi:10.1073/pnas.0703267104
2. Tobias Overath et al., 'The cortical analysis of speech-specific temporal structure revealed by responses to sound quilts', *Nature Neuroscience*, vol. 18, issue 6, 2015, pp. 903–11, doi:10.1038/nn.4021
   It has long been known that specific regions of the brain act together in producing speech. This research indicates that one area is used exclusively to analyse the sounds of human speech, but not other sounds—even when the speech sounds are mixed up from different languages.
3. David Maximiliano Gómez et al., 'Language universals at birth', *Proceedings of the National Academy of Sciences*, vol. 111, no. 16, 2014, pp. 5837–41, www.pnas.org/cgi/doi/10.1073/pnas.1318261111
4. Josef Parvizi et al., 'Electrical stimulation of human fusiform face-selective region distorts face perception', *The Journal of Neuroscience*, vol. 32, no. 43, 2012, pp. 14915–20.
5. Sara J. Shettleworth, 'Memory in food-hoarding birds', *Scientific American*, vol. 248, no. 3, 1983, pp.86–94.
6. Robin Dunbar, 'Neocortex size as a constraint on group size in primates', *Journal of Human Evolution*, vol. 20, no. 6, 1992, pp. 469–93, doi:10.1016/0047-2484(92)90081-J

7. Colin A. Chapman and J.A. Teichroeb, 'What influences the size of groups in which primates choose to live?' *Nature Education Knowledge,* vol. 3, no. 10, 2012.
8. Robin Dunbar, *Grooming, Gossip, and the Evolution of Language,* Harvard University Press, Cambridge, MA, 1998.
9. Frank W. Marlowe, 'Hunter-gatherers and human evolution', *Evolutionary Anthropology*, vol. 14, no. 2, 2005, pp. 54 –67, doi:10.1002/evan.20046
10. Tim Ingold, *The Appropriation of Nature: Essays on Human Ecology and Social Relations,* University of Iowa Press, Iowa City, 1987, pp. 130–64.
11. Carla Aimé et al., 'Human genetic data reveal contrasting demographic patterns between sedentary and nomadic populations that predate the emergence of farming', *Molecular Biology and Evolution*, vol. 30, issue 12, 2013, pp. 2629–44, doi:10.1093/molbev/mst156
12. Jean-Pierre Bocquet-Appel, 'Estimates of upper palaeolithic meta-population size in Europe from archaeological data', *Journal of Archaeological Science,* vol. 32, issue 11, 2005, pp. 1656–68.
13. Guillaume Chapron et al., 'Recovery of large carnivores in Europe's modern human-dominated landscapes', *Science,* vol. 346, no. 6216, 2014, pp. 1517–19, doi:10.1126/science.1257553
14. Joseph W. Eaton and A.J. Mayer, *Man's Capacity to Reproduce: The Demography of a Unique Population,* The Free Press, Glencoe, Illinois,1954.
15. Gregory Clark, *A Farewell to Alms: A Brief Economic History of the World*, Princeton University Press, Princeton and Oxford, 2007, pp. 22–3.
16. Peter T. Ellison, *On Fertile Ground: A Natural History of Human Reproduction,* Harvard University Press, Cambridge, MA, 2001, pp. 83–126.
17. Ibid., pp. 119–20.
18. R.S. Meindl, 'Human populations before agriculture', in *The Cambridge Encyclopaedia of Human Evolution,* eds. S Jones, R. Martin and D. Pilbeam, pp. 406–10. Cambridge University Press, Cambridge, 1992.

19. John. C. Caldwell and B.K. Caldwell, 'Pretransitional population control and equilibrium', *Population Studies,* vol. 57, no. 2, 2003, pp. 199–216.
20. John. C. Caldwell and B.K. Caldwell, 'Was there a neolithic mortality crisis?' *Journal of Population Research,* vol. 20, issue 2, 2003, pp. 153–68.

Chapter 3 Climate

1. A useful diagram of Earth's energy budget has been produced by NASA. It can be viewed and downloaded as a poster from the website. Accessed 18 Nov. 2015, http://science-edu.larc.nasa.gov/energy_budget/
2. Christopher J. Poulsen et al., 'Long-term climate forcing by atmospheric oxygen concentrations', *Science,* vol. 348, no. 6240, 2015, pp. 1238–41, doi:10.1126/science.1260670
3. James Lovelock, *Gaia: A New Look at Life on Earth,* Oxford University Press, Oxford, 1979 (3rd edition, 2000).
4 In April, 2014, when James Lovelock was 94, the British Science Museum opened an exhibition celebrating his life and career. Roger Highfield, reporting in *The Telegraph,* described Lovelock as 'the most famous and influential independent scientist of the 20th century'.
5. Vaclav Smil, *Energy: A Beginner's Guide,* Oneworld Publications, Oxford, 2006, p. 14. Vaclav Smil is a recognized authority on the subject and this book offers an easily accessible and comprehensive explanation of energy—ranging from how nature works to the use of energy in modern societies.
6. The website of the US National Oceanic and Atmospheric Administration (NOAA) provides a description of the global conveyor belt, complete with animations. Accessed 26 Nov. 2015, http://oceanservice.noaa.gov/education/kits/ currents/06conveyor2.html
7 Anon., 'Ocean circulation is changing, and we need to know why', *Nature,* Editorial, vol. 556, issue 7700, 2018, p 149
8. Quirin Schiermeier, 'Oceans under surveillance', *Nature,* vol. 494, 2013, pp. 167–8, doi:1038/497167a
9. U.S. Geological Survey, 'Volcanic Gases and Climate Change Overview'. Accessed 20, Nov. 2015, http://volcanoes.usgs.gov/hazards/gas/climate.php

10. Greg O'Hare, J. Sweeney and R. Wilby, *Weather, Climate and Climate Change: Human Perspectives*, Routledge, London and New York, 2013, pp. 226–9.
11. Spencer Weart, *The Discovery of Global Warming (*Revised Edition), Harvard University Press, Cambridge, MA, 2008. In his acclaimed book, Spencer Weart discusses chaos theory in the context of climate change.

Chapter 4 People through the Pleistocene

1. Ian McDougall et al., 'Stratigraphic placement and age of modern humans from Kibish, Ethiopia', *Nature*, vol. 433, issue 7027, 2005, p. 733, doi:10.1038/nature03258
2. Jean-Jacques Hublin et al., 'New fossils from Jebel Irhoud, Morocco and the pan-African origin of *Homo sapiens', Nature*, vol. 546, issue 7657, 2017, pp. 289–92.
3. Definitions are based on proposals by the International Commission on Stratigraphy (ICS) that were ratified by the International Union for Geological Sciences (IUGS) in 2009.
4. David E. Anderson, A.S. Goudie and A.G. Parker, *Global Environments through the Quaternary: exploring environmental change,* 2nd edn, 2013, Oxford University Press, Oxford.
5. Anders E. Carlson, 'Ice sheets and sea level in earth's past', *Nature education Knowledge,* vol. 3, no. 10, 2011. Accessed 28 Nov. 2015, http://www.nature.com/scitable/knowledge/library/ice-sheets-and-sea-level-in-earth-24148940
6. Mark A. Maslin et al., 'East African climate pulses and early human evolution', *Quaternary Science Reviews,* vol. 101, 2014, pp 1–17, doi:10.1016/j.quascirev.2014.06.012
7. Bernard A. Wood, 'Fifty years after *Homo habilis*', *Nature*, vol. 508, issue 7494, 2014, pp. 31–3, doi:10.1038/508031a
8. Kathy D. Schick and N.P. Toth, *Making Silent Stones Speak: Human Evolution And The Dawn Of Technology*. Simon & Schuster, New York, 1993, pp 135–42.
9. Nicholas P. Toth et al., 'Pan the tool-maker: investigations into the stone tool-making and tool-using capabilities of a bonobo *(Pan paniscus)', Journal of Archaeological Science,* vol. 20, issue 1, 1993, pp. 81–91.

10. Chris B. Stringer, 'Evolution of early humans', in *The Cambridge Encyclopedia of Human Evolution*, eds. S Jones, R. Martin and D. Pilbeam, Cambridge University Press,Cambridge,1992, pp. 241–51.
11. Joseph V. Ferraro et al., 'Earliest archaeological evidence of persistent hominin carnivory', *PLoS One,* vol. 8, no. 4, 2013, doi:10.1371/journal.pone.0062174
12. Francesco Berna et al., 'Microstratigraphic evidence of *in situ* fire in the Acheulean strata of Wonderwerk Cave, Northern Cape province, South Africa', *Proceedings of the National Academy of Sciences* , vol. 109, no. 20, 2012, pp. 7593–4, doi:10.1073/pnas.1117620109
13. Christopher J. Lepre, 'An earlier origin for the Acheulean', *Nature,* vol. 477, issue 7362, 2011, pp. 82–5, doi:10.1038/nature10372
14. John A.J. Gowlett, 'Tools—the palaeolithic record', in *The Cambridge Encyclopedia of Human Evolution*, pp. 350–60.
15. Margaret E. Lewis and L. Werdelin, 'Carnivoran dispersal out of Africa during the early Pleistocene: relevance for hominins?', in *Out of Africa I: The First Hominin Colonization of Eurasia,* eds. J.G. Fleagle et al., Springer, Dortrecht, 2010, pp. 13–26.
16. Anon. 'The formation of the modern Essex landscape: the Ipswichian Interglacial', website of *GeoEssex: Promoting the geology and landscape of Essex.* Accessed 20 Nov. 2015, http://www.geoessex.org.uk/the_formation_of_the_modern_essex_landscape.html
17. Gisela H. Kopp et al., 'Out of Africa, but how and when? The case of hamadryas baboons *(Papio hamadryas)', Journal of Human Evolution,* vol. 76, 2014, pp. 154–64, doi:10.1016/j.jhevol.2014.08.003
18 Ian Tattersall, *Becoming Human: Evolution and Human Uniqueness,* Harcourt Brace, Orlando, FL, 1998, pp. 136–40.
19. Thor J. Bergman, J.E. Phillips-Conroy and C.J. Jolly, 'Behavioral variation and reproductive success of male baboons (*Papio anubis* x *Papio hamadryas*) in a hybrid social group', *American Journal of Primatology,* vol. 70, no. *2,* 2008, pp. 136–47.
20. Reid Ferring et al., 'Earliest human occupations at Dmanisi (Georgian Caucasus) dated to 1.85–1.78 Ma', *Proceedings of the National Academy of Sciences* , vol. 108, no. 26, 2011, pp 10432–36, doi:10.1073/pnas.1106638108

21. David Lordkipanidze et al., 'A complete skull from Dmanisi, Georgia, and the evolutionary biology of early *Homo*', *Science,* vol. 342, no. 6156, 2013, pp. 326–31, doi:10.1126/science.1238484
22. Alan G. Thorne and M.H. Wolpoff, 'The multiregional evolution of humans', *Scientific American,* vol. 266, no. 4, 1992, pp. 76–83, doi:10.1038/scientificamerican0492-76
23. Alan C.Wilson and R.L. Cann, 'The recent African genesis of humans', *Scientific American,* vol. 266, no. 4, 1992, pp. 68–73, doi:10.1038/scientificamerican0492-66
This, and the previous reference, explain the two rival models of modern human origins at the popular science level.
24. Ann Gibbons, 'Who were the Denisovans?', *Science,* vol. 333, no. 6046, 2011, pp. 1057–88.
25. Matthias Meyer et al., 'A high-coverage genome sequence from an archaic Denisovan individual', *Science,* vol. 338, no. 6104, 2012, pp. 222–26, doi:10.1126/science.1224344
26. Peter J. Obendorf, C.E. Oxnard and B.J. Kefford, 'Are the small human-like fossils found on Flores human endemic cretins?', *Proceedings of the Royal Society B*, vol. 275, issue 1640, 2008, pp. 1287–96, doi:10.1098/rspb.2007.1488
27. Chris Stringer, 'Small remains still pose big problems', *Nature,* vol. 514, issue 7523, 2014, pp. 427–9.
28. Julia Galway-Witham and C. Stringer, 'How did *Homo sapiens* evolve?', *Science,* vol. 360, no. 6395, 2018, pp. 1296–8.
29. Matthias Meyer et al., 'A high Coverage Genome Sequence'.
30. Zenobia Jacobs et al., 'Ages for the Middle Stone Age of southern Africa: implications for human behavior and dispersal', *Science,* vol. 322, no. 5902, 2008, pp. 733–5, doi:10.1126/science.1162219
31. Kyle S. Brown et al., 'An early and enduring advanced technology originating 71,000 years ago in South Africa', *Nature,* vol. 491, issue 7425, 2012, pp. 590–3, doi:10.1038/nature11660
32. Bruce A. Buck, 'Ancient technology in contemporary surgery', *The Western Journal of Medicine*, vol. 136, no. 3, 1982, pp. 265–9.
33. Michael J. Scott and M.J. Scott, 'Obsidian surgical blades: modern use of a Stone Age implement', *The Journal of Dermatologic Surgery and Oncology,* vol. 8, issue 12, 1982, pp. 1050–2, doi:10.1111/j.1524-4725.1982.tb01093.x
34. Andrew Sherratt, 2005, 'Obsidian Trade in the Near East, 14,000 to 6500 BC', website of *ArchAtlas*.

35. Ian C. Gilby et al., 'No evidence of short-term exchange of meat for sex among chimpanzees', *Journal of Human Evolution*, vol. 59, no. 1, 2010, p. 44, doi:10.1016/j.jhevol.2010.02.006

Chapter 5 A hesitant transition

1. William Amos and J.I. Hoffman, 'Evidence that two main bottleneck events shaped modern human genetic diversity', *Proceedings of the Royal Society B*, vol. 277, issue 1678, pp. 131–7, doi:10.1098/rspb.2009.1473
2. Marilyn Menotti-Raymond and S.J. O'Brien, 'Dating the genetic bottleneck of the African cheetah', *Proceedings of the National Academy of Sciences*, vol. 90, no. 8, pp. 3172–6, doi:10.1073/pnas.90.8.3172
3. Anna Revedin et al., 'Thirty thousand-year-old evidence of plant food processing', *Proceedings of the National Academy of Sciences*, vol. 107, no. 44, 2010, pp. 18815–9, www.pnas.org/cgi/doi/10.1073/pnas.1006993107
4. Chip Walter, 'Origins of Art' *National Geographic*, January 2015. This is an interesting and beautifully-illustrated account for non-specialists.
5. William Rendu et al., 'Evidence supporting an intentional Neandertal burial at La Chapelle-aux-Saints', *Proceedings of the National Academy of Sciences*, vol. 111, no. 1, pp. 81–6, www.pnas.org/cgi/doi/10.1073/pnas.1316780110
6. Nicholas J. Conard, M. Malina and S.C. Münze, 'New flutes document the earliest musical tradition in southwestern Germany', *Nature*, vol. 460, issue 7256, 2009, pp. 737–40, doi:10.1038/nature08169
7. M. Aubert et al., 'Pleistocene cave art from Sulawesi, Indonesia', *Nature*, vol.514, issue 7521, 2014, pp. 223–7, doi:10.1038/nature13422
8. Josephine C.A. Joordens et al., '*Homo erectus* at Trinil on Java used shells for tool production and engraving', *Nature*, vol. 518, issue 7538, 2015, pp. 228–231, doi:10.1038/nature13962
9. Chip Walter, 'Origins of Art'.
10. Lisa A. Maher, T. Richter and J.T. Stock, 'The pre-Natufian epipaleolithic: long-term behavioral trends in the Levant'. *Evolutionary Anthropology*, vol. 21, issue 2, pp. 69–81.

11. Michel Louis Séfériadès, 'Spondylus and long-distance trade in prehistoric Europe', in *The Lost World of Old Europe: The Danube Valley, 5000-3500 BC,* ed. David W. Anthony, Princeton University Press, New Jersey, 2010, pp. 179–91.
12. Paul S. Martin, 'The Discovery of America', *Science*, vol. 179, no. 4077, 1973, pp. 969–74.
13. Todd A. Surovell et al., 'Test of Martin's overkill hypothesis using radiocarbon dates on extinct megafauna', *Proceedings of the National Academy of Sciences,* vol. 113, no. 4, 2016, pp.86–91, doi:10.1073/pnas.1504020112
14. Website of the U.S. National Oceanic and Atmospheric Administration (NOAA), National Climatic Data Center.
15. Intergovernmental Panel on Climate Change, *Working Group 1 (The Physical Science Basis), Contribution to Fifth Assessment Report,* (AR5), 2013.
16. Lewis R. Binford, *Constructing Frames of Reference: An Analytical Method for Archaeological Theory Building Using Ethnographic and Environmental Data Sets*, University of California Press, Berkeley, 2001. Lewis Binford (1931–2011) was one of the most influential archaeologists of the twentieth century. In this book he brings his scientific approach to hunting and gathering.
17. G.W.T.A. Groot Bruinderink, E. Hazebroek and H. Van Der Voot, 'Diet and condition of wild boar, *Sus scrofu scrofu*, without supplementary feeding' *Journal of Zoology*, vol. 233, issue 4, 1994, pp. 631–48.
18. Eric Baubet, C. Bonenfant and S. Brandt, 'Diet of the wild boar in the French Alps', *Galemys,* vol. 16, 2004, pp. 101–13.
19. Claudia Melis, et al., 'Biogeographical variation in the population density of wild boar (*Sus scrofa*) in western Eurasia', *Journal of Biogeography,* vol. 33, issue 5, 2006, pp. 803–11, doi:10.1111/j.1365-2699.2006.01434.x
20. Vernon L. Smith, 'Humankind in prehistory: economy, ecology, and institutions' in *The Political Economy of Customs and Culture,* eds. T. L. Anderson and R. Simmons, Rowman & Littlefield, Lanham, MD, 1993, pp. 157–84.
21. Garrett Hardin, 'The tragedy of the commons', *Science,* vol. 162, issue 3859, 1968, pp. 1243–8, doi:10.1126/science.162.3859.1243
22. Kent V. Flannery, 'Origins and ecological effects of early domestication in Iran and the Near East', in *The Domestication and*

Exploitation of Plants and Animals, eds. P.J. Ucko and G.W. Dimbleby, Aldine, Chicago, IL, 1969, pp. 73–100.

23. Mary C. Stiner, N.D. Munro and T.A. Surovell, 'The tortoise and the hare: small-game use, the broad-spectrum revolution, and paleolithic demography, *Current Anthropology,* vol. 41, no. 1, 2000, pp. 39–73.
24. Mary C. Stiner, 'Thirty years on: the "broad spectrum revolution" and paleolithic demography', *Proceedings of the National Academy of Sciences,* vol. 19, no. 13, 2001, pp. 6993–6.
25. Natalie D. Munro, 'Epipaleolithic subsistence intensification in the southern Levant: the faunal evidence', in *The Evolution of Hominin Diets: Integrating Approaches to the Study of Palaeolithic Subsistence*, eds. J. Hublin and M.P. Richards, Springer, Dordrecht, 2009, pp. 141–55.
26. Ehud Weiss et al., 'The broad spectrum revisited: evidence from plant remains', *Proceedings of the National Academy of Sciences* ,vol. 101, no. 26, 2004, pp. 9551–5.
27. Marco Madella et al., 'The exploitation of plant resources by Neanderthals in Amud Cave (Israel): the evidence from phytolith studies', *Journal of Archaeological Science,* no. 29, 2002, pp. 703–19.
28. Lisa A. Maher, T. Richter and Jay T. Stock, The pre-Natufian epipaleolithic, p. 69.
29. Tobias Richter and L.A. Maher, 'Terminology, process and change: reflections on the epipalaeolithic of south-west Asia', *Levant*, vol. 45, issue 2, 2013, pp. 121–32.
30. Ofer Bar-Yosef, 'The Natufian culture in the Levant, threshold to the origins of agriculture', *Evolutionary Anthropology,* vol. 6, issue 5, 1998, p. 161.

31 Ibid., p. 174.

32. Ibid., p. 159.
33. Katharina Neumann, 'The romance of farming: plant cultivation and domestication in Africa', in A.B. Stahl, ed. *African Archaeology: A Critical Introduction,* Blackwell, Oxford, 2004, pp. 249–75.
34. Dorian Q. Fuller, 'Agricultural origins and frontiers in south Asia: a working synthesis', *Journal of World Prehistory,* vol. 20, no. 2, 2006, doi:10.1007/s10963-006-9006-8
35. Peter J. Richerson, R. Boyd and R.L. Bettinger, 'Was agriculture impossible during the Pleistocene but mandatory during the Holocene? A climate change hypothesis', *American Antiquity*, vol. 66,

no. 3, 2001, pp. 387-411. Accessed 25 Oct. 2015, http://www.jstor.org/stable/2694241

36. Ehud Weiss and D. Zohary, 'The neolithic southwest Asian founder crops: their biology and archaeobotany', *Current Anthropology,* vol. 52, supplement 4, 2011, pp. 237–54.
37. Frank Hole, 'Origins of agriculture' in *The Cambridge Encyclopedia of Human Evolution*, eds. S Jones, R. Martin and D. Pilbeam, Cambridge University Press, Cambridge, 1992, pp. 373–79.
38. Jared Diamond and P. Bellwood, 'Farmers and their languages: the first expansions', *Science*, vol. 300, no. 5619, 2003, pp. 597–603, doi:10.1126/science.1078208
39. Peter J. Richerson, R. Boyd and R.L. Bettinger, 'Was agriculture impossible, p. 389.
40. Alan H. Simmons, *The Neolithic Revolution in the Near East: Transforming the Human Landscape,* University of Arizona Press, Tucson, 2011, p. 28.
41. Solange Rigaud, F. d'Errico and M. Vanhaeren, 'Ornaments reveal resistance of north European cultures to the spread of farming', *PLoS One*, vol 10, no. 4, 2015, doi:10.1371/journal.pone.0121166
42. Vernon L. Smith, 'Humankind in prehistory', p.174.
43. Peter J. Richerson, R. Boyd and R.L. Bettinger, 'Was agriculture impossible', p. 395.

Chapter 6 Farmers, towns and temples

1. Robert J. Ruben, 'A time frame of critical/sensitive periods of language development', *International Journal of Otolaryngology and Head & Neck Surgery,* vol. 51, no. 3, 1999, pp. 85–9.
2. Ann Gibbons, 'Neandertal genes linked to modern diseases', *Science* vol. 351, no. 6274, 2016, pp. 648–9.
3. By convention, 'before present' is taken to mean before 1950, by which year carbon dating had been established.
4. Guy Bar-Oza, M. Zederb, and F. Hole, 'Role of mass-kill hunting strategies in the extirpation of Persian gazelle (Gazella subgutturosa) in the northern Levant', *Proceedings of the National Academy of Sciences,* vol. 108, no. 18, 2011, pp. 7345–50, www.pnas.org/cgi/doi/10.1073/pnas.1017647108 .

5. A.M.T. Moore, G.C. Hillman and A.J. Legge, *Village on the Euphrates: from Foraging to Farming at Abu Hureyra,* Oxford University Press, New York, 2000.
6. A.M.T. Moore, 'The impact of accelerator dating at the early village of Abu Hureyra on the Euphrates' *Radiocarbon*, vol. 34, no. 3, 1992, pp. 850–8.
7. Michael Mann, *The Sources of Social Power, Volume 1, a history of power from the beginning to AD 760,* Cambridge University Press, Cambridge, 1986, p. 34.
8. V. Gordon Childe, 'The urban revolution', *The Town Planning Review,* vol. 21, no. 1, 1950, pp. 3–17. Accessed 29 Feb. 2016, http://www.jstor.org/stable/40102108

   Childe had been developing and publishing his ideas since the 1930s but he liked to reach the ordinary reader. This article, published by the Liverpool University Press, was exceptionally widely-read.
9. Bleda S. Düring, 'The anatomy of a prehistoric community: reconsidering Çatalhöyük', in *From Prehistoric Villages to Cities: settlement aggregation and community transformation,* ed. Jennifer Birch, Routledge, New York, 2013, pp. 23–43.
10. Ian Hodder, *Religion in the Emergence of Civilization: Çatalhöyük as a Case Study,* Cambridge University Press, New York, 2010.
11. ibid., p. 17.
12. UNESCO World Heritage, *Neolithic Site of Çatalhöyük,* Accessed 29 Feb. 2016, http://whc.unesco.org/en/list/1405
13. Robert A. Carter and G. Philip, 'Deconstructing the Ubaid' in *Beyond the Ubaid: Transformation and Integration in the Late Prehistoric Societies of the Middle East,* eds. R. A. Carter and G. Philip, Oriental Institute of the University of Chicago, Chicago, 2010, pp. 1–22.
14. Ibid., p.6.
15. Nicole Brisch, 'History and chronology' in *The Sumerian World*, ed. H. Crawford, Routledge, New York, 2013, p.111.
16. Jon Taylor, 'Administrators and scholars: the first scribes' in *The Sumerian World*, ed. H. Crawford, Routledge, New York, 2013, p. 297.
17. Irving Finkel and J. Taylor, *Cuneiform,* British Museum Press, London, 2015.

18. Nicole Brisch, 'History and chronology' in *The Sumerian World*, p.111.
19. Ibid.
20. Petr Charvát, *Mesopotamia Before History*, Routledge, New York, 2003, p.159.
21. Susan Pollock, 'Politics of food in early Mesopotamian societies', *Origini*, XXXIV, 2012, pp. 153–68.
22. Susan Pollock, *Ancient Mesopotamia: The Eden that Never Was*, Cambridge University Press, Cambridge, 1999, p. 47.
23. Gwendolyn Leick, *Mesopotamia: the Invention of the City*, Penguin, London, 2002, pp. 39–40.
24 Ibid., p. 41.
25. Gil J. Stein 'Producers, patrons and prestige: Craft specialists and emergent elites in Mesopotamia from 5500–3100 B.C.', in *Craft Specialization and Social Evolution: in Memory of V. Gordon Childe*, ed. B. Wailes, University of Pennsylvania Press, Philadelphia, 1996, p. 32.
26. Kent V. Flannery and J. Marcus, *The Creation of Inequality: How our Prehistoric Ancestors Set the Stage for Monarchy, Slavery, and Empire*, Harvard University Press, Cambridge, MA, 2014, p. 454.
27. Susan Pollock, 'Politics of food in early Mesopotamian societies', p. 161.
28. Guillermo Algaze, *Ancient Mesopotamia at the Dawn of Civilization: The Evolution of an Urban Landscape*, University of Chicago Press, Chicago, 2008, p. 66.
29. Ibid., p. 95.
30. Heather Lechtman, 'Arsenic bronze: dirty copper or chosen alloy? A view from the Americas', *Journal of Field Archaeology*, vol. 23, no. 4, 1996, pp. 477–514.
31. Susan Pollock, *Ancient Mesopotamia*, p. 9.

Chapter 7 Towns to city states

1. Ron Pinhasi and J.T. Stock, *Human Bioarchaeology of the Transition to Agriculture*, John Wiley, Chichestor, 2011.
2. Herman Pontzer et al., 'Hunter-gatherer energetics and human obesity', *PLoS One*, vol 7, no. 7, 2012, doi:10.1371/journal.pone.0040503

3. United Nations agencies (WHO and FAO) have established 1,800 kilocalories a day as the minimum adult energy intake for basic metabolism. Accessed 15 March 2016, http://www.un.org/en/globalissues/briefingpapers/food/whatishunger.shtml
4. Anna Rooseveld, 'The history of mortality and physiological stress', in *Paleopathology at the Origins of Agriculture*, eds M.N. Cohen and G.J. Armelagos, Academic Press, Cambridge, MA 1984, pp. 243–4.
5. Ron Pinhasi and J.T. Stock, *Human Bioarchaeology of the Transition to Agriculture.*
6. Ian Amit et al., 'Difference in death? A lost inhumation neolithic cemetery with Britain's earliest case of rickets, at Balevullin, western Scotland', *Proceedings of the Prehistoric Society,* vol. 81, 2015, pp. 199–214, doi:http://dx.doi.org/10.1017/ppr.2015.7
7. Christine J. Adler et al., 'Sequencing ancient calcified dental plaque shows changes in oral microbiota with dietary shifts of the neolithic and industrial revolutions', *Nature Genetics,* vol. 45, issue 4, 2013, pp. 450–55, doi:10.1038/ng.2536
8 Nathan D. Wolfe, C.P. Dunavan and J. Diamond, 'Origins of major human infectious diseases', *Nature,* vol. 447, issue 7142, 2007, pp. 279–283, doi:10.1038/nature05775
9 Ethne Barnes, *Diseases and Human Evolution,* University of New Mexico Press, Albuquerque, 2005, p. 69.
10 Alvin Powell, 'Evolution in real time: 59,000 generations of bacteria, plus freezer, yield startling results', *Harvard Gazette,* Feb. 13, 2014. Accessed 14 May 2016, doi:http://news.harvard.edu/gazette/story/2014/02/evolution-in-real-time/
11. Nathan D. Wolfe, C.P. Dunavan and J. Diamond, 'Origins of major human infectious diseases'.
12. Ibid.
13. Simon T. Powers and L. Lehmann, 'An evolutionary model explaining the neolithic transition from egalitarianism to leadership and despotism', *Proceedings of the Royal Society B,* vol. 281, issue 1791, 2014, doi:10.1098/rspb.2014.1349
14. Martha Vicinus and B. Nergaard, *Ever Yours, Florence Nightingale: Selected Letters,* Virago Press, London, 1989, p. 200.
15. Neha Mahajan and A. Woodward, 'Seven-month-old infants selectively reproduce the goals of animate but not inanimate agents',

Infancy, vol. 14, no. 6, 2009, pp. 667–79, doi:10.1080/15250000903265184

16. R. Saxe, J.B. Tenenbaum and S. Carey, 'Secret agents: inferences about hidden causes by 10- and 12-month-old infants', *Psychological Science*, vol. 16, no. 12, 2005, pp. 995–1001.
17. Edward B. Tylor, *Anthropology: an Introduction to the Study of Man and Civilization,* Appleton and Co. New York, 1896, p. 343.
18. Kelly Bulkeley, *Big Dreams: the Science of Dreaming and the Origins of Religion*, Oxford University Press, New York, 2016.

Chapter 8 Economics, human nature, and early limits to growth

1. Mark Thornton, 'The Mystery of Adam Smith's Invisible Hand Resolved', New Working Paper, *Mises Institute*, 2006. Accessed 21 Apr. 2016, https://mises.org/blog/new-working-paper-smiths-invisible-hand
2. Adam Smith, *An Inquiry Into the Nature and Causes of the Wealth of Nations,* Eleventh London Edition, volume 1, Oliver D. Cooke, Hartford, CT, 1811, p. 10. This printing has been digitized by Google.
3. Sarah F. Brosnan and M.J. Beran, 'Trading behavior between conspecifics in chimpanzees, *Pan troglodytes*', *Journal of Comparative Psychology*, vol 123, no. 2, 2009, pp. 181–94. http://dx.doi.org/10.1037/a0015092
4. Adam Smith, *An Inquiry Into the Nature and Causes of the Wealth of Nations*, p. 46.
5. Fred Block and M.R. Somers, '*The Power of Market Fundamentalism: Karl Polanyi's Critique*', Harvard University Press, Cambridge, MA, 2014.
6. Karl Polanyi, *The Great Transformation: The Political and Economic Origins of Our Time,* Beacon Press, Boston, Mass. 2001. This edition includes a *Forward* by Joseph Stiglitz and an *Introduction* by Fred Block.
7. Ibid., pp. 45–6.
8. Ibid., p. 55.
9 Ibid., p. 58.
10. Fred Block, *Introduction* in Karl Polanyi, *The Great Transformation.*

11. Mark Granovetter, 'Economic action and social structure: the problem of embeddedness', *American Journal of Sociology*, vol. 91, issue 3, 1985, pp. 481–510.

12 Li Xing, 'Understanding China's economic success: "embeddedness" with Chinese characteristics', *Asian Culture and History*; vol. 8, no. 2, 2016, pp. 18–31, doi:10.5539/ach.v8n2p18

13. Morris Silver, *Economic Structures of Antiquity*, Greenwood Press, Westport, CT, 1995.

14. Susan Pollock, *Ancient Mesopotamia: the Eden that Never Was*, Cambridge University Press, Cambridge, 1999.

15. Marc Van De Mieroop, 'The invention of interest', in *The Origins of Value: The financial Innovations that Created Modern Capital Markets,* eds. W.N. Goetzmann and K.G. Rouwenhorst, Oxford University Press, New York, 2005, p. 18.

16. Morris Silver, *Economic Structures of Antiquity*, p. 203.

17. Ibid., p. 11.

18. Walter Scheidel, (ed) *The Cambridge Companion to the Roman Economy*, Cambridge University Press, Cambridge, 2012, Chapter 1. This was first published as 'Approaching the Roman economy', *Princeton/Stanford Working Papers in Classics*, 2010. Accessed 17 May, 2016, https://www.princeton.edu/~pswpc/pdfs/scheidel/091007.pdf

19. Richard Wolff, *A Pocket Dictionary of Greek and Roman Gods and Goddesses*, British Museum Press, London, 2003.

20. *Encyclopædia Britannica.* Accessed 14, July 2016: https://www.britannica.com/biography/Odoacer

21. Bamber Gascoigne, 'History of banking', *History World.* Accessed 16 Sept. 2016, www.historyworld.net/wrldhis/PlainTextHistories.asp?historyid=ac19

22. Bank of England, 'A brief history of banknotes'. Accessed 16 Sept. 2016, www.bankofengland.co.uk/banknotes/Pages/about/history.aspx

23. Massimo Livi-Bacci, *A Concise History of World Population,* 5th edn, Wiley-Blackwell, Chichester, 2012, p. 25.

24. Ole J. Benedictow, *The Black Death, 1346–1353: the Complete History*, Boydell Press, Woodbridge, 2004.

25. Massimo Livi-Bacci, *A Concise History,* p. 41.

26. S. Kathleen Lyons et al., 'Holocene shifts in the assembly of plant and animal communities implicate human impacts', *Nature*, vol. 529, issue 7584, 2016, pp. 80–3.
27. Barry K. Gills and W.R. Thomson, *Globalization and Global History*, Routledge, London and New York, 2006, p. 168.
28. Ibid., p.166.
29. Vaclav Smil, *Energy: A Beginner's Guide*, Oneworld, Oxford, 2006, pp. 87–8.

Chapter 9 The coming of industry

1. Manchester City Council, *Spinning the Web: The story of the Cotton Industry*. Accessed 20 June, 2016, http://www.spinningtheweb.org.uk/
2. John Oakland, *Contemporary Britain: A Survey With Texts*, Routledge, London and New York, 2002, p 5.
3. John Noble Wilford, 'Site in Turkey yields oldest cloth ever found', *New York Times*. Accessed 23 May, 2016, http://www.nytimes.com/1993/07/13/science/site-in-turkey-yields-oldest-cloth-ever-found.html?pagewanted=all
4 Eliso Kavavadze et al., '30,000 year-old wild flax fibres', *Science*, vol. 325, no. 5946, 2009, p. 1359, doi:10.1126/science.1175404
5. E.J.W. Barber, *Prehistoric Textiles: The Development of Cloth in the Neolithic and Bronze Ages*, Princeton University Press, New Jersey, 1991, pp. 24–5.
6. ABC News, 'Wool world record: 40kg fleece shorn off overgrown sheep in Canberra, RSPCA says'. Accessed 25 May, 2016, http://www.abc.net.au/news/2015-09-03/wool-world-record-set-after-42kg-fleece-shorn-off-canberra-sheep/6746200
7. E.J.W. Barber, *Prehistoric Textiles*, pp. 4, 31–2.
8. Ravi Srivastava, 'Changing employment conditions of the Indian workforce and implications for decent work', *Global Labour Journal*, vol. 3, no. 1, 2012, doi:http://dx.doi.org/10.15173/glj.v3i1.1113
9. R.S. Fitton, *The Arkwrights, Spinners of Fortune*, Manchester University Press, Manchester, 1989, p.152. This work is particularly detailed for the purpose of historical record.
10. Wikepedia, 'History of the steam engine'. Article presenting an excellent summary of the development of the steam engine.

Accessed 21 June, 2016.
https://en.wikipedia.org/wiki/History_of_the_steam_engine

11. Paul Markillie, 'The third industrial revolution', *The Economist*, Apr. 21–27, 2012, accessed 19 May, 2016, http://www.economist.com/node/21553017
12. Klaus Schwab, 'The fourth industrial revolution: what it means, and how to respond', *World Economic Forum*. Accessed 19 May, 2016, https://www.weforum.org/agenda/2016/01/the-fourth-industrial-revolution-what-it-means-and-how-to-respond/
13. Beatrijs Sterk, 'The rise and structural change of textile production in Europe', *European textile network*. Accessed 27 May, 2016, http://etn-net.org/routes/intro/indust_intro_GB.htm
14. *Encyclopædia Britannica*, 'Industrial Revolution'. Accessed 3 June, 2016, http://www.britannica.com/event/Industrial-Revolution
15. Ian Haddow, 'When UK GDP last outstripped the US', *BBC News*, 7 Jan. 2008. Accessed 3 June, 2016, http://news.bbc.co.uk/2/hi/uk_news/7174996.stm
16. Peter Duus, 'Introduction' in *The Cambridge History of Japan, Vol. 6, The Twentieth Century*, ed. P. Duus, Cambridge University Press, Cambridge, 1988, pp. 14–5.
17. Gregory Clark, *A Farewell to Alms: a Brief Economic History of the World*, Princeton University Press, Princeton and Oxford, 2007, pp. 41–2.
18. Thomas Malthus, *An Essay on the Principle of Population*, London 1798, Chapter 2. Published in 1998, by Electronic Scholarly Publishing Project, http://www.esp.org
19. Ibid., Chapter 10.
20. Ibid., Chapter 6.
21. Robert Dorfman, 'Thomas Robert Malthus and David Ricardo', *Journal of Economic Perspectives*, vol. 3, no. 3, 1989, pp. 153–64.
22. Karl Marx, *Capital Vol. 1*, Penguin classics, London, 1990, pp. 783–4.
23. Julian Huxley, *Soviet Genetics and World Science*, Chatto & Windus, London, 1949, p. 38.
24. Massimo Livi-Bacci, *A Concise History of World Population,* 5th edn, Wiley-Blackwell, Chichester, 2012, pp. 25, 90.
25. Chelsea Wald, 'The secret history of ancient toilets', *Nature*, vol. 533, issue 7604, pp. 456–8, doi:10.1038/533456a

26. Massimo Livi-Bacci, *A Concise History of World Population,* p. 25.
27. Vaclav Smil, *Energy: A Beginner's Guide*, Oneworld, Oxford, 2006, pp. 107–8.
28. Jed O. Kaplan, K.M. Krumhardt and N. Zimmermann, 'The prehistoric and preindustrial deforestation of Europe', *Quaternary Science Reviews,* vol. 28, 2009, pp. 3016–34.

Chapter 10 Into a different world

1. *Encyclopædia Britannica*, 'Dark Ages'. Accessed 18 May, 2016, https://www.britannica.com/event/Dark-Ages
2. From *The History of Four-footed Beasts and Serpents* by Edward Topsell, London, 1607. The extracts are from sections entitled 'Of the Weasel', and 'Of the vulgar little Mouse', from the website of *Early English Books Online.* Accessed 18 Aug. 2015, http://quod.lib.umich.edu/e/eebo/A42668.0001.001?
   Topsell was a clergyman who recorded what he understood to be facts. His books were circulated by respected and influential people of the time so were widely considered to be authoritative.
3. John A. Moore, *Science as a Way of Knowing: The Foundations of Modern Biology.* Harvard University Press, Cambridge, MA, 1993, p. 66.
4. Hillel Ofek, 'Why the Arabic world turned away from science', *The New Atlantis*, no. 30, 2011, pp. 3–23.
5. Ann B. Shteir, *Cultivating Women, Cultivating Science: Flora's Daughters and Botany in England, 1760–1860,* Johns Hopkins University Press, Baltimore, 1996.
6. Samantha George, 'Linnaeus in letters and the cultivation of the female mind: botany in an English dress', *Journal for Eighteenth Century Studies*, vol. 28, issue 1, 2005, pp. 1–18.
7. University of California Museum of Paleontology, 'Jean-Baptiste Lamarck (1744–1829)'. Accessed 24 July, 2016, http://www.ucmp.berkeley.edu/history/lamarck.html
8. Charles Darwin, *The Origin of Species by Means of Natural Selection*, Penguin, London, 1968, p. 54.
9. University of California Museum of Paleontology, 'Jean-Baptiste Lamarck'.
10. A Wallace Memorial Fund was started in 1999, initially for the purpose of restoring Wallace's grave in Dorset, UK. The Fund now maintains a comprehensive website with information about his life

and work, including an illustrated list of medals and other honours awarded to him. Accessed 21 Aug. 2016, http://wallacefund.info/honours-wallace-received

11. Darwin Correspondence Project, Cambridge University Library, University of Cambridge. Accessed 27 July, 2016, https://www.darwinproject.ac.uk/
12. J.W. Burrow, 'Editor's Introduction' in Darwin's *Origin of Species*, Penguin, 1968, p. 34.
13. Alfred Russel Wallace, *My Life: A Record of Events and Opinions*, Vol. 1, Chapman & Hall, London, 1905, p. 365. Accessed 25 July, 2016, http://Wallace-online.org/
14. Arnold C. Brackman, *A Delicate Arrangement: The Strange Case of Charles Darwin and Alfred Russel Wallace*, Times Books, New York, 1980.
15. Alfred Russel Wallace, *Darwinism: an Exposition of the Theory of Natural Selection with Some of its Applications*, Macmillan, London & New York, 1889.
16. Alfred Russel Wallace, *My Life*, Vol. 1, pp. 361–3.
17. Darwin's personal recollections, written mainly for his own records, and for his children, were collected, edited by his son, Francis, and published by John Murray in 1887 as *The Autobiography of Charles Darwin*. There have been more recent versions, including one edited by his granddaughter, Nora Barlow, and published by Collins in 1958. The reference to Malthus appears in the section entitled 'My several publications'.
18. Charles Darwin, *The Origin of Species*, p. 458.
19. Terence Ball, *Reappraising Political Theory*, Oxford University Press, Oxford, 1995, pp. 237–41.
20. It is sometimes claimed that Spencer has been treated unfairly by those who quote him out of context. This may be so but on the subject of social evolution Spencer made his views quite clear in his book *Social Statics* (1851). Then, in *The Man Versus the State* (1892) he declared that Darwin's work had further strengthened his arguments (p.110, American edition, Caxton Printers, Idaho, 1940).
21. Gavan Tredoux maintains a website with the intention of making freely available everything that Galton wrote—often in facsimile. Accessed 20 Aug. 2016, http://galton.org/
22. John A. Moore, *Science as a Way of Knowing*, pp. 258–9.
23. Ibid., pp. 279–80.

24. Philip K. Wilson, 'Eugenics' *Encyclopædia Britannica.* A dispassionate summary of the eugenics movement accessed 1 Aug. 2016, https://www.britannica.com/science/eugenics-genetics

25. The evolution of cooperation is notoriously difficult to explain in cases where individuals would need to sacrifice themselves for the good of the group (i.e. group selection). The Royal Society organized two symposia on the topic, with papers published in a theme issue of *Philosophical Transactions of the Royal Society B*, vol 371, issue 1687, February 2016, http://rstb.royalsocietypublishing.org/content/371/1687

26. John Ray's *The Wisdom of God Manifested in the Works of His Creation* (1691) was the basis of Paley's thinking in *Natural Theology*, published in 1802.

27. Julian Huxley, *Evolution: the Modern Synthesis*, has been produced in several editions, the most recent being published by MIT Press, Cambridge, MA, 2009.

Chapter 11 Post-industrial economics and ecology

1. Following the Anglo-Nepalese war of 1814–16, some territory controlled by Nepal was ceded to the British East India Company. That territory was never returned to Nepal and is now held by India. Ethiopia, then Abyssinia, was occupied by Italy from 1936 to 1941.

2. Anne O. Krueger, in the Annual Gilbert Lecture, Rochester University, New York, Apr. 6, 2006. Accessed 22 Aug. 2016. https://www.imf.org/en/News/Articles/2015/09/28/04/53/ sp040606

3. A.G. Kenwood and A.L. Lougheed, *The Growth of the International Economy 1820–2000: an Introductory Text*, 4th edn. Routledge, London, 1999, pp. 61, 79.

4. Anne O. Krueger, in the Annual Gilbert Lecture.

5. Julia Driver, 'The history of utilitarianism', *The Stanford Encyclopedia of Philosophy*, accessed 22 Aug. 2016, http://plato.stanford.edu/archives/win2014/entries/utilitarianism-history/

6. John Stuart Mill, *Utilitarianism*, Parker, Son and Bourn, London, 1863. Accessed in facsimile, 22 Aug. 2016, https://archive.org/details/a592840000milluoft

7. John Stuart Mill, *Principles of Political Economy*, D. Appleton & Company, New York, 1885. Accessed online 11 Aug. 2016, https://www.gutenberg.org/files/30107/30107-pdf.pdf

8. Ellen Frankel Paul, 'Laissez faire in nineteenth-century Britain: fact or myth?', *Literature of Liberty*. Vol. iii, no. 4, 1980, pp. 5–38. Accessed 22 Aug. 2016 from *Library of Economics and Liberty*, ttp://www.econlib.org/library/Essays/LtrLbrty/fplLNB.html
9. C. Wilfred Jenks and J. Schregle, 'Labour law', *Encyclopædia Britannica*, accessed 22 Aug. 2016, https://www.britannica.com/topic/labour-law
10. Some historians follow the Marxist historian, Eric Hobsbaum, in referring to the period from the French revolution to World War 1 as the 'long 19th century'.
11. Harry Magdoff, C.E. Nowell and R.A. Webster, 'European expansion since 1763', *Encyclopædia Britannica.* Accessed 9 Aug. 2016, https://www.britannica.com/topic/colonialism/ European-expansion-since-1763
12. Charles Darwin, *The Origin of Species*, p. 125.
13. United Nations Population Division, Department of Economic and Social Affairs, *The World at Six Billion.* Accessed 17 Aug. 2016, http://www.un.org/esa/population/publications/sixbillion/sixbilcover.pdf
14. The ecological footprint is put forward as a measure of the natural resources that humans demand from nature. This is seen in relation to the capacity of nature to provide the resources and to absorb the waste. Global Footprint Network: http://www.footprintnetwork.org/our-work/
15. Dániel Csaba Lánczi, 'Practice of sustainability in an eco village: ecological footprint of Krishna Valley in Hungary', published online by *The Global Ecovillage Network.*
16. Sixty years after the London smog of 1952, *The Daily Mail, Australia* published a well-illustrated article on the event. Accessed 23 Aug. 2016, http://www.dailymail.co.uk/news/article-2243732/Pea-souper-killed-12-000-So-black-screen-cinemas-So-suffocatingly-lethal-ran-coffins-How-Great-Smog-choked-London-60-years-ago-week.html
17. The National Oceanic and Atmospheric Administration (NOAA), Ocean Service Education website. Accessed 23 Aug. 2016, http://oceanservice.noaa.gov/education/kits/ pollution/02history.html
18. The Bentham Project at University College London is an exceptional source of information on Jeremy Bentham. See: https://web.archive.org/web/20070103005132/http://www.ucl.ac.uk/Bentham-Project/info/project.htm

19. Laxman Singh Kandari et al., 'Conservation and management of sacred groves, myths and beliefs of tribal communities: a case study from north India', *Environmental Systems Research*, Dec. 2014, vol. 3, no,16, doi:10.1186/s40068-014-0016-8

20. Robin W. Doughty, *Feather Fashions and Bird Preservation: A Study in Nature Protection*, University of California Press, Berkeley, 1975.

Chapter 12 Living it up and crashing down

1. Joshua Zeitz, 'The roaring twenties', *The Gilder Lehrman Institute of American History*, accessed 13, Sept. 2016, www.gilderlehrman.org/history-by-era/roaring-twenties/essays/roaring-twenties

2. The phrase 'the war to end war', or a variant of it, became a catchphrase in 1914. In that year, the author H.G. Wells had a book published with the title *The War That Will End War*. The American president, Woodrow Wilson, later reached a wider audience when he said 'This is a war to end all wars.'

3. In summary, the conflict was between the Central Powers and the Allied Powers, in which the Central Powers were Austria-Hungary, Germany, Bulgaria and the Ottoman Empire. The main allies were Serbia, Russia, France, the British Empire, Italy, Belgium and the United States.

4. Eric Johnston, 'Japan and the Great War', *Japan times*, 16 June, 2014. Accessed 2 Oct. 2016, www.japantimes.co.jp/news/2014/06/16/reference/japans-gambit-wwi-set-stage-dark-future/#.V9Jm__l96_4

5. From the homepage of *U.S Army Europe*, accessed 26 Sept.2016, www.eur.army.mil/organization/history.html

6 Hugh Rockoff, 'U.S. Economy in World War I', *EH.Net Encyclopedia*, edited by Robert Whaples. Accessed 8 Sept. 2016, http://eh.net/encyclopedia/u-s-economy-in-world-war-i/

7. Richard Lambert, 'Crashes, bangs and wallops' *Financial Times*, 13 July 2008.

8. Robert W. Fogel, 'Simon S. Kuznets', *Biographical Memoirs V.79,* The National Academies Press, Washington, D.C., 2001, pp. 203-30. Accessed 7 July 2017, https://www.nap.edu/read/ 10169/chapter/13

9. John R. Walter, 'Depression-era bank failures: the great contagion or the great shakeout?', *Economic Quarterly*, Reserve Bank of Richmond, vol. 91, no. 1, 2005, pp. 39–54.

10. 'Unemployment statistics during the great depression', *United States History*. Accessed 30 Sept. 2016, www.u-s-history.com/pages/h1528.html
11. A.G. Kenwood and A.L. Lougheed, *The Growth of the International Economy 1820–2000*, 4th edn, Routledge, London, 1999, pp 187–90.
12 Milton Friedman and A.J. Schwartz, *The Great Contraction 1929–1933*, Princeton University Press, Princeton, 2008. This title was originally published in 1965 and the new edition includes comments by Ben Bernanke. The text is mainly an extract from Friedman & Schwartz' classic of 1963: *A Monetary History of the United States*, 1867–1960.
13. The quip was made by Dorothy Parker, as reported in David Greenberg's *Calvin Coolidge*, Times Books, New York, 2006, p. 9.
14. Dalmer D. Hoskins, 'U.S. social security at 75 years: an international perspective', *Social Security Bulletin*, vol. 70, no. 3, 2010. Accessed 22 Sept. 2016, https://www.ssa.gov/policy/docs/ssb/v70n3/v70n3p79.html
15. *Encyclopaedia Britannica*, 'New deal', accessed 11 Sept. 2016, https://www.britannica.com/event/New-Deal
16. Philip Scranton, 'War reparations and the great depression: echoes', *Bloomberg View*, accessed 30 Aug. 2016, https://www.bloomberg.com/view/articles/2011-12-05/war-reparations-and-the-great-depression-echoes
17. Stuart D. Goldman, 'The forgotten Soviet-Japanese war of 1939', *The Diplomat*, 28 Aug. 2012. Accessed 3 Oct. 2016, http://thediplomat.com/2012/08/the-forgotten-soviet-japanese-war-of-1939/
18. John Lukacs, W.F. Knapp and A. Bullock, 'Adolf Hitler, dictator of Germany', *Encyclopaedia Britannica*, accessed 3 Oct. 2016, https://www.britannica.com/biography/Adolf-Hitler

Chapter 13 Starting afresh

1. Ian Buruma, *Year Zero: a History of 1945,* Penguin, New York, 2013.
2. Number from the website of the National WWII Museum, New Orleans. Accessed 3 Oct. 2016, https://www.nationalww2museum.org/students-teachers/student-resources/research-starters/research-starters-worldwide-deaths-world-war

3. Richard P. Rumelt, 'World War ll stimulus and the postwar boom', *Wall Street Journal*, 30 July, 2011. Accessed Oct. 2016, https://www.wsj.com/articles/SB10001424053111903554904576458413656841844
4. Paul Krugman, 'Franklin Delano Obama?', *New York Times*, 10 Nov. 2008. Accessed 4 Oct. 2016, http://www.nytimes.com/2008/11/10/opinion/10krugman.html
5. Christopher J. Tassava, 'The American economy during World War II', *Economic History Association*, 10 Feb. 2008, (online).
6. 'Lend-lease and military aid to the allies in the early years of World War II' *Office of the Historian*, U.S. Dept. of State, accessed 6 Oct. 2016, https://history.state.gov/milestones/1937-1945/lend-lease
7. Robert Costanza et al., 'A short history of GDP: moving towards better measures of human well-being', *Solutions*, vol 5, issue 1, 2014, pp. 91–7.
8. World Trade Organization, 'Understanding the WTO - The GATT years: from Havana to Marrakesh'. Accessed 15 Oct. 2017, https://www.wto.org/english/thewto_e/whatis_e/tif_e/ fact4_e.htm The last of the trade rounds, the Uruguay round that finally led to the WTO, went on for 7 years.
9. A.G. Kenwood and A.L. Lougheed, *The Growth of the International Economy 1820–2000*, 4th edn., Routledge, London, 1999, pp. 242–5.
10. Mark Harrison, 'The Soviet Union after 1945: economic recovery and political repression', *Past and Present*, vol. 210 (Supplement 6), 2011, pp. 103–20.
11. David R. Henderson, 'German economic miracle' *The Concise Encyclopedia of Economics*, accessed 8 Oct. 2016, http://www.econlib.org/library/Enc/GermanEconomicMiracle.html
12. 'Occupation and reconstruction of Japan, 1945–52', *Office of the Historian*, U.S. Dept. of State. Accessed 6 Oct. 2016, https://history.state.gov/milestones/1945-52/ japan-reconstruction
13. Eric Johnston, 'Japan's gambit in WWI set stage for a dark future', *Japan Times*. Accessed 9 Sept. 2016, http://www.japantimes.co.jp/news/2014/06/16/reference/japans-gambit-wwi-set-stage-dark-future/#.V9Jm__l96_4
14. J.D.C. Bennett, 'Medical advances consequent to the Great War 1914-1918', *Journal of the Royal Society of Medicine*, vol. 83, no. 11, 1990, pp. 738–42 Accessed 1 Oct. 2016, www.ncbi.nlm.nih.gov/pmc/articles/PMC1292926/

15. Glenn Beall, 'By design: World War II, plastics, and NPE', *Plastics Today, Newsletter*, 9 Apr. 2009.
16. United Nations, Department of Economic and Social Affairs, Population Division, *World Population Prospects: The 2017 Revision*, Key findings, p. 2, https://esa.un.org/unpd/wpp/Publications/Files/WPP2017_KeyFindings.pdf
17. A.G. Kenwood and A.L. Lougheed, *The Growth of the International Economy 1820–2000*, p. 247.
18. Andrew Glyn et al., 'Rise and fall of the golden age' in *The Golden Age of Capitalism: Reinterpreting the Postwar Experience*, eds. S.A. Marglin and J.B Schor, Clarendon Press, Oxford, 1990, reprinted 2007, pp. 39–125.
19 Michael Corbett, 'Oil shock of 1973–4' *Federal Reserve History Gateway*, accessed 3 Nov. 2016, http://www.federalreservehistory.org/Events/DetailView/36
20. A.G. Kenwood and A.L. Lougheed, *The Growth of the International Economy 1820–2000*, pp 247, 249.
21. Angus Maddison, 'The world economy in the second half of the twentieth century', in *The World Economy: Vol 1: A Millennial Perspective*, OECD Publishing, Paris, 2006, p. 125.
22. *World Factbook*, 'GDP - composition by sector of origin', Accessed 3 Nov. 2016, www.cia.gov/library/publications/the-world-factbook/fields/2012.html
23. Commission on Growth and Development, *The Growth Report: Strategies for Sustained Growth and Inclusive Development*, World Bank, Washington, D.C., 2008, accessed 16 Oct. 2017, https://openknowledge.worldbank.org/handle/10986/6507
24. Ibid.
25. Encyclopædia Britannica, 'John Maynard Keynes'.
26. John Maynard Keynes, *The Economic Consequences of the Peace*, Harcourt, Brace and Howe, New York, 1920.
27. John McWhorter, 'When people were proud to call themselves neoliberal', *The Atlantic*, 30 May, 2017. Accessed 16 Oct. 017, https://www.theatlantic.com/business/archive/2017/05/history-of-neoliberal-meaning/528276/
28. Samuel Brittan, 'Milton Friedman, economist, dies aged 94', *Financial Times*, Nov 16, 2006.
29. The Economist, 'Where does the buck stop?', *The Economist,* 13 Aug. 2016.

Chapter 14 The rise of environmentalism

1. Swedish Environmental Protection Agency, 'Sweden's Environment Problems and Protection: 1960–2010', accessed 3 Nov. 2016 https://www.naturvardsverket.se/Documents/publikationer6400/978-91-620-8501-8.pdf?pid=4183
2. Rachel Carson, *Silent Spring*, Houghton Mifflin, Boston, 1962. The book is reported to have sold more than two million copies. An anniversary edition was published by Houghton Mifflin in 2002.
3. Extension Toxicology Network, 'DDT (dichlorodiphenyltrichloroethane)'. Accessed 15 Nov. 2017, http://pmep.cce.cornell.edu/profiles/extoxnet/carbaryl-dicrotophos/ ddt-ext.html
4. K.G. Drouillard, 'Biomagnification.' in *Ecotoxicology*, eds. S.E Jorgensen and B.D. Fath, Academic Press, Amsterdam, 2010, pp. 92–6. The Clear Lake example is taken from M.L Flint and R. van den Bosch, *Introduction to Integrated Pest Management,* Plenum Press, New York, 1981.
5. Glen A. Fox, 'Eggshell quality: its ecological and physiological significance in a DDE-contaminated common tern population', *The Wilson Bulletin* vol. 88, no. 3, Sept. 1976, pp. 459–77. It is well known that DDT and its metabolites can cause eggshells to be thinner and more easily broken during incubation. But an eggshell is not just a calcium container; it is structured to permit oxygen and CO_2 exchange within critical limits. Early studies, such as the one cited above, showed that DDE can affect shell development in various ways that prove fatal to bird embryos without the shell being broken.
6. Iman Al-Saleh et al., 'Levels of DDT and its metabolites in placenta, maternal and cord blood and their potential influence on neonatal anthropometric measures.', *Science of the Total Environment,* vol. 416, February 2012, pp. 62–74, doi:10.1016/j.scitotenv.211.11.020
7. Eliza Grizwold, 'How *Silent Spring* ignited the environmental movement', *New York Times Sunday Magazine*, 23 Sept. 2012.
8. The Stockholm Convention on Persistent Organic Pollutants, Stockholm, May 2001, accessed 23 Nov. 2016, http://chm.pops.int/Portals/0/Repository/convention_text/UNEP-POPS-COP-CONVTEXT-FULL.English.PDF
9. Alliance of Religions and Conservation (ARC), *The Assisi Declarations: Messages on Humanity and Nature from Buddhism, Christianity, Hinduism, Islam & Judaism*, Assisi, 1986. Accessed

2 Dec. 2016, http://www.arcworld.org/downloads/the%20assisi%20declarations.pdf

10. David A. Johnson, 'The wide-ranging role of the microbiome', *Medscape*, 15 Sept. 2015, accessed 18 Nov. 2016, http://www.medscape.com/viewarticle/850764#vp_3
11. Holmes Rolston III, 'Value in nature and the nature of value', in *Philosophy and the Natural Environment*, eds. R. Attfield and A. Belsey, Cambridge University Press, Cambridge, 1994, pp. 13–30.
12 Extracted from the Mission Statement of animal rights advocates Gary L. Francione and Anna Charlton, as it appears on their website: *Animal Rights: the Abolitionist Approach.* Accessed 2 Dec. 2016, http://www.abolitionistapproach.com/about/the-six-principles-of-the-abolitionist-approach-to-animal-rights/#.WEDpudV96iQ
13. Alan Drengson, 'Ecophilosophy, ecosophy and deep ecology: an overview.', *Ecocentrism.* Accessed 2 Dec. 2016, http://www.ecospherics.net/pages/DrengEcophil.html
14. Murray Bookchin, 'Social ecology versus deep ecology: a challenge for the ecology movement', first published in 1987 and subsequently posted on *Anarchy Archives*, accessed 1 Dec. 2016, http://dwardmac.pitzer.edu/Anarchist_Archives/bookchin/socecovdeepeco.html
15. Jeffrey M. Jones, 'Americans' identification as "environmentalists" down to 42%', *Gallup Social Issues*, 22 Apr. accessed 17 Dec. 2016, http://www.gallup.com/poll/190916/americans-identification-environmentalists-down.aspx
16. Brendan O'Neill, 'There is nothing left-wing about environmentalism.' A talk broadcast by the Australian Broadcasting Commission on Radio National's *Counterpoint* in 2012. Transcript accessed 30 Nov. 2016, http://www.abc.net.au/news/2012-09-26/oneill---environmentalism/4280782

Chapter 15 Beyond pesticides: global responses

1. World Environmental Organization, http://www.world.org/weo/top1000
2. Philippe Sands et al., *Principles of International Environmental Law*, Third Edition, Cambridge University Press, Cambridge, 2012, p. 22.
3. Chris Hails, 'WWF from 1961 to 2006: WWF's approach to conservation from its inception to 2006'. Accessed 26 Nov. 2016,

www.panda.org/who_we_are/history/wwwf_conservation_1961_2006/

4. IUCN has an extensive website: ttps://www.iucn.org/about
5. For an overview of the *Red List*, see: http://www.iucnredlist.org/about/overview
6. Charity Navigator, 'Giving statistics', accessed 28 Nov. 2016, http://www.charitynavigator.org/index.cfm/bay/content.view/cpid/42
7. The first *Living Planet Report* was published in 1998. The report for 2018, can be found at: https://www.worldwildlife.org/pages/living-planet-report-2018
8. The Conference was held at UNESCO House in Paris from 4–13 Sept. 1968. The final report is entitled *Intergovernmental conference of experts on the scientific basis for rational use and conservation of the resources of the biosphere.* Available from: http://unesdoc.unesco.org/images/0001/000172/017269eb.pdf
9. *Report of the United Nations Conference on the Human Environment,* Stockholm, 5–16 June 1972. Available from: http://www.un-documents.net/aconf48-14r1.pdf
10. The illegal trade in wildlife is estimated to take about US$ 19 billion a year but the value of an illegal international trade is obviously difficult to estimate accurately. See CITES, https://www.cites.org/eng
11. World Commission on Environment and Development, *Our Common Future*, Oxford University Press, Oxford, 1987, http://www.un-documents.net/our-common-future.pdf
12. Ibid., p. 1.
13. Ibid., p. 8.
14. Philippe Sands et al., *Principles of International Environmental Law*, p. 42.
15 The 6 categories are: Ia Strict Nature Reserve; Ib Wilderness Area; II National Park; III Natural Monument or Feature; IV Habitat/Species Management Area; V Protected Landscape/Seascape. See: https://www.iucn.org/theme/protected-areas/about/protected-area-categories
16. Philippe Sands et al., *Principles of International Environmental Law*, p. 49.
17. Jeff Tollefson and N. Gilbert, 'Earth Summit: Rio report card' *Nature*, vol. 486, issue 7401, 2012, pp. 20–3.

18. Roland Geyer, J.R. Jambeck and K.L. Law, 'Production, use, and fate of all plastics ever made' *Science Advances*, vol. 3, no. 7, e1700782 2017, doi:10.1126/sciadv.1700782
19. L. Lebreton et al., 'Evidence that the Great Pacific Garbage Patch is rapidly accumulating plastic', *Scientific Reports*, vol. 8, article no. number: 4666, 2018, https://doi.org/10.1038/s41598-018-22939-w
20. Stephanie L. Wright and F.J. Kelly, 'Plastic and human health: a micro issue?', *Environmental Science &. Technology*, vol 51, no.12, 2017, pp 6634–47, doi: 10.1021/acs.est.7b00423
21. Anon., 'Microplastics and human health—an urgent problem', *The Lancet Planetary Health,* Editorial, vol. 1, no. 7 e254-e297, doi:10.1016/S2542-5196(17)30121-3.
22. United Nations Convention on Biodiversity, objectives from Article 1. See: https://www.cbd.int/doc/legal/cbd-en.pdf
23. Secretariat of the Convention on Biological Diversity, *Sustaining Life on Earth: How the Convention on Biological Diversity promotes nature and human well-being,* Apr. 2000, https://www.cbd.int/doc/publications/cbd-sustain-en.pdf
24. Adam Vaughan, 'Public awareness of the biodiversity crisis is virtually non-existent', Environment Blog, *The Guardian.* Accessed 1 Dec. 2017, https://www.theguardian.com/environment/blog/2010/sep/16/ public-awareness-biodiversity-crisis
25. Nick M. Haddad et al., 'Habitat fragmentation and its lasting impact on Earth's ecosystems', *Science Advances*, vol. 1, no. 2, doi:10.1126/sciadv.1500052
26. J.P. Rodríguez et al., 'Globilization of conservation: a view from the south', Science, vol. 317, no. 5839, 2007, pp. 755–6, doi:10.1126/science.1145560
27. FAO, *Bushmeat Hunting.* Online note accessed 1 Dec. 2017, http://www.fao.org/forestry/wildlife/67287/en/
28. William J. Ripple, et al., 'Bushmeat hunting and extinction risk to the world's mammals' *Royal Society Open Science.* **3**: 160498, 2016, http://dx.doi.org/10.1098/rsos.160498
29. WWF, *Living Planet Report 1998,* pp. 1–4, accessed 13 Dec. 2016, http://awsassets.panda.org/downloads/lpr_living_planet_report_1998.pdf
30. WWF, *Living Planet Report 2016: Risk and Resilience in a New Era,* WWF International, Gland, Switzerland, p.77. Accessed 13 Dec. 2016, http://www.worldwildlife.org/pages/ living-planet-report-2016

31. Linus Blomqvist et al., 'Does the shoe fit? real versus imagined ecological footprints', *PLOS Biol,* 11 (11) 2013, doi.org/10.1371/journal.pbio.1001700

32. Andrea Saltelli and Mario Giampietro, 'Footprints to nowhere', *Ecological Indicators*, vol 46, 2014, pp. 610–21, doi:10.1016/j.ecolind.2014.01.030

Chapter 16 Climate steals the thunder

1. G.S. Callendar, 'The artificial production of carbon dioxide and its influence on temperature', *Quarterly Journal of the Royal Meteorological Society*, 1938, doi:10.1002/qj.49706427503
2. A comprehensive website is maintained at: https://www.ipcc.ch/
3. IPCC, *Climate change 2014: synthesis report. Contribution of Working Groups I, II and III to the Fifth Assessment Report of the Intergovernmental Panel on Climate Change*, IPCC, Geneva, Switzerland, 2015, pp. 2, 4.
4. J.R. Porter et al., 'Food security and food production systems', in: *Climate Change 2014: Impacts, Adaptation, and Vulnerability, Part A: Global and Sectoral Aspects. Contribution of Working Group II to the Fifth Assessment Report of the Intergovernmental Panel on Climate Change,* IPCC, Geneva, Switzerland, 2014.
5. FAO, *The State of Food and Agriculture: Climate Change, Agriculture and Food Security*, FAO, Rome, 2016.
6. FAO, *The Future of Food and Agriculture – Trends and Challenges,* FAO, Rome, 2017.
7. David Robert Grimes, 'Impartial journalism is laudable. But false balance is dangerous', *The Guardian,* 9 Nov. 2016, accessed 27 Dec. 2016, https://www.theguardian.com/science/blog/2016/nov/08/impartial-journalism-is-laudable-but-false-balance-is-dangerous
8. Naomi Oreskes and E.M. Conway, *Merchants of Doubt: how a handful of scientists obscured the truth on issues from tobacco smoke to global warming,* Bloomsbury Press, New York, 2010.
9. Craig D. Idso, R.M. Carter and S.F. Singer, *Why Scientists Disagree About Global Warming: The NIPCC Report on Scientific Consensus,* The Heartland Institute, Arlington Heights, Ill, 2015, pp. 38–47. Accessed 2 Jan. 2017, https://www.heartland.org/publications-resources/publications/why-scientists-disagree-about-global-warming

10. James Lovelock, *The Vanishing Face of Gaia*, Penguin, London, 2009, p. 8. Lovelock here refers to a climate scientist's description of political manipulations during an IPCC working group.
11. E.A.G. Schuur et al., 'Climate change and the permafrost carbon feedback', *Nature,* vol. 520, issue 7546, 2015, pp. 171–9, doi:10.1038/nature14338
12. Richard Black, 'Hijacked by climate change?', accessed 2 Jan. 2017, http://news.bbc.co.uk/2/hi/science/nature/8223611.stm
The programme, *The Great Climate Change Hijack,* went to air on BBC Radio Four in the UK on Thursday 27 August, 2009.
13. Ibid.
14. Michael Le Page, 'Was Kyoto climate deal a success? Figures reveal mixed results.' *New Scientist Daily News*, 14 June 2016, accessed 2 Jan. 2017, https://www.newscientist.com/article/ 2093579-was-kyoto-climate-deal-a-success-figures-reveal-mixed-results/
15. Anon., 'Sulfur dioxide from volcanoes keeping the earth cool', *Nature World News*, 5 Mar. 2013, accessed 20 Sep. 2015, http://www.natureworldnews.com/articles/731/20130305/sulfur-dioxide-volcanoes-keeping-earth-cool.htm
16. Reto Knutti et al., 'A scientific critique of the two-degree climate change target', *Nature Geoscience*, vol 9, issue 1, 2016, pp. 13–8.
17. Both the BP *Statistical Review* and *Energy Outlook* publications are available in pdf format from the BP website, https://www.bp.com/
18. Reto Knutti et al., 'A scientific critique of the two-degree climate change target'.

Chapter 17 Population

1. Steven Hecht Orzach et al., 'The human sex ratio from conception to birth', *Proceedings of the National Academy of Sciences*, vol. 112, no. 16, 2015, doi:10.1073/pnas.1416546112
2. Arthur Haupt, T.T. Kane and C. Haub, *PRB'S Population Handbook*, 6th edn, Population Reference Bureau, Washington D.C., 2011, p. 13.
3. Ibid., p. 6.
4. Dudley Kirk, 'Demographic transition theory', *Population Studies*, vol. 50, 1996, p. 384.

5. Martin W. Holdgate and G.D Ness, 'Hopes for the future' in *Population and Global Security*, ed. N. Polunin, Cambridge University Press, Cambridge, 1998, pp. 271–97.
6. David M. Cutler, A.S. Deaton and A.L. Muney, 'The determinants of mortality', Working Paper 11963, National Bureau of Economic Research, Cambridge, MA, 2006. Accessed 4 May 2017, http://www.nber.org/papers/w11963
7. Alastair Leithead, 'Can Niger break out of its cycle of poverty?', *BBC News*, accessed 9 Nov. 2017, http://www.bbc.com/news/world-africa-41018081
8. David E. Bloom and D.L. Luca, 'The global demography of ageing: facts, explanations, future', *Working Paper no. 130*, Program on the Global Demography of Aging at Harvard University, accessed 23 Apr. 2016, http://www.hsph.harvard.edu/pgda/working/
9. Joydeep Battacharya and S. Chakraborty, 'Contraception and the fertility transition', *Economics Working Papers* (2002–2016) no. 29, Iowa State University, 2014. Accessed I May 2018, http://lib.dr.iastate.edu/econ_las_workingpapers/29
10. John Cleland et al., 'Family planning: the unfinished agenda', *The Lancet*, vol. 368, no. 9549, 2006, pp. 1810–27, doi:http://dx.doi.org/10.1016/S0140-6736(06)69480-4
11. Martha M. Campbell, N. Prata and M. Potts, 'The impact of freedom on fertility decline', *Journal of Family Planning and Reproductive Health Care*, vol. 39, 2012, pp 44–50, doi:10.1136/jfprhc-2012-100405
12. Eleanor Randall, 'Family planning programmes review', *Population Matters*, London, 2012. Accessed 3 May 2017, http://www.populationmatters.org/documents/programmes_eview.pdf
13. John Bongaarts and S.W. Sinding, 'A response to critics of family planning programs' *International Perspectives on Sexual and Reproductive Health*, Vol. 35, no. 1, 2009, pp. 39–44.
14. United Nations, Dept of Economic and Social Affairs, *World Population Prospects: The 2010 Revision, Vol. 1, Comprehensive Tables*. New York, 2011.
15. United Nations, Dept of Economic and Social Affairs, *World Population Prospects: The 2017 Revision, Key Findings and Advance Tables*, New York, 2017.
16. Population Reference Bureau, *2016 World Population Data Sheet,* Population Reference Bureau, Washington D.C., p.10.

17. United Nations, Dept of Economic and Social Affairs, *World Population Prospects: The 2017 Revision.*

18. *Ibid.*

19. National Research Council, *Beyond Six Billion: Forecasting the World's population,* eds. J. Bongaarts and R. A. Bulatao, National Academy Press, Washington, D.C., 2000. Accessed 1 May 2018, https://www.nap.edu/read/9828/chapter/4#38

20. United Nations, Dept of Economic and Social Affairs, *World Urbanization Prospects: The 2014 Revision, New York, 2014.* Accessed 3 May 2017, https://esa.un.org/unpd/wup/publications/files/wup2014-highlights.Pdf

21. David Dodman et al., 'African urbanisation and urbanism: implications for risk accumulation and reduction', *International Journal of Disaster Risk Reduction*, vol. 26, 2017, pp. 7–15, https://doi.org/10.1016/j.ijdrr.2017.06.029

22. FAO, *Urban and peri-urban forestry in Africa: the outlook for woodfuel*, FAO Working Paper, Rome 2012. Accessed 2 Apr. 2018, http://www.fao.org/docrep/016/i1973e/i1973e00.pdf

23. This topic is thoroughly explored in *The Oxford Handbook of Work and Aging*, eds. J.W. Hedge and W.C. Borman, Oxford University Press, New York, 2012.

24. International Labour Organization, Summary of the *World Employment and Social Outlook 2016: Trends for Youth,* ILO, Geneva. Accessed 12 Nov 2017, https://www.bibb.de/dokumente/pdf/world_employment_social_outlook_youth.pdf

25. International Labour Organization, *World Employment and Social Outlook: Trends 2017,* ILO, Geneva. Accessed 21 Sept. 2017, http://www.ilo.org/wcmsp5/groups/public/---dgreports/---dcomm/---publ/documents/publication/wcms_541211.pdf

26. Jan Van Bavel and D.S. Reher, 'The Baby Boom and Its Causes: What We Know and What We Need to Know', *Population and Development Review*, vol. 39, no. 2, 2013, pp. 257–88, doi:10.1111/j.1728-4457.2013.00591.x

27. Joshua R. Goldstein, 'How populations age', in *International Handbook of Population Aging*, ed. P. Uhlenberg, Springer, Dordrecht, 2009, pp. 7–18.

28. United Nations, Dept of Economic and Social Affairs, *World Population Ageing 1950-2050*, pp. 6–7, accessed 16 Apr. 2017, http://www.un.org/esa/population/publications/worldageing19502050/

29. Ibid.
30. U.S. Census Bureau, *65+ in the United States: 2010,* U.S. Government Printing Office, Washington, D.C., 2014. Accessed 16 Apr. 2017, https://www.census.gov/content/dam/Census/library/publications/2014/demo/p23-212.pdf
31. United Nations, Dept of Economic and Social Affairs, *World Population Ageing 1950-2050,* p. 20.
32. The remarks by Italy's health minister were widely reported. For example: Fred Pierce, 'Baby slump puts Italy at risk of dying' *New Scientist Daily News,* 16 Feb. 2015, accessed 29 Apr. 2017, https://www.newscientist.com/article/dn26982-baby-slump-puts-italy-at-risk-of-dying/
33. Pew Research Center, *Attitudes About Aging: a Global Perspective,* 2014, accessed 5 May 2017, http://www.pewglobal.org/2014/01/30/attitudes-about-aging-a-global-perspective/
34. Jinil Kim, 'The effects of demographic change on GDP growth in OECD economies' *IFDP Notes,* Federal Reserve System, Washington D.C., 28 Sept. 2016.
35. Anon. 'Procreation day delivers results', *The Washington Times,* August 15, 2007, accessed 5 May 2017, http://www.washingtontimes.com/news/2007/aug/15/procreation-day-delivers-results
36. Gaia Pianigiani, 'Italy's fertility day call to make babies arouses anger, not ardor, *The New York Times,* 13 Sept. 2016.
37. Olivier Thévenon, 'The influence of family policies on fertility in France: lessons from the past and prospects for the future', in *Low Fertility, Institutions, and their Policies: Variations Across Industrialized Countries,* Springer, Cham, Switzerland, 2016, pp. 49–76.
38. Anthony A. Noce, D. Qirjo and N. De Silva, 'enticing the stork: can we evaluate pro-natal policies before having children?', *Economic Affairs,* Vol. 36, issue 2, 2016, pp. 184–202.
39. OECD, 'Ideal and actual number of children', OECD Social Policy Division, accessed 14 May 2017, https://www.oecd.org/els/family/SF_2_2-Ideal-actual-number-children.pdf
40. Joshua R. Goldstein, 'How populations age', in *International Handbook of Population Aging,* p. 15.
41. United Nations, *International Migration Report 2017,* Dept of Economic and Social Affairs, New York, 2017.

42. Joshua R. Goldstein, 'How populations age', in *International Handbook of Population Aging*, p.16

43. Taina Leinonena et al., 'Health as a predictor of early retirement before and after introduction of a flexible statutory pension age in Finland', *Social Science & Medicine*, Vol. 158, June 2016, pp.149–57, https://doi.org/10.1016/j.socscimed.2016.04.029

44. Sarah Harper, *Ageing societies: Myths, Challenges and Opportunities*, Routledge, New York, 2013, pp. 21–2.

45. Warren C. Sanderson and S. Scherbov, 'Remeasuring Aging', *Science,* vol. 329, issue 5997, 2010, pp. 1287–8.

46. Katina Kardamanidis et al., 'Hospital costs of older people in New South Wales in the last year of life', *Medical Journal of Australia*, vol. 187, no. 7, 2007, pp. 383–6.

47. Warren C. Sanderson and S. Scherbov, 'Remeasuring Aging', pp. 1287–8.

48 OECD Data, 'Life expectancy at 65' accessed 26 Apr. 2016, https://data.oecd.org/healthstat/life-expectancy-at-65.htm

49. Warren C. Sanderson and S. Scherbov, 'A new perspective on population aging', *Demographic Research*, vol. 6, no. 2, 2007, pp. 27–58, doi:10.4054/DemRes.2007.16.2

50. EU Newsroom, 'The EU Covenant on Demographic Change steps in for age-friendly Europe', 29 Jan. 2016.

Chapter 18 Numbers and the green revolution

1. Jason Riggio et al., 'The size of savannah Africa: a lion's (*Panthera leo*) view', *Biodiversity and Conservation*, vol. 22, issue 1, 2013, pp. 17–35, doi:10.1007/s10531-012-0381-4

2. Paul R. Ehrlich and J.P. Holdren, 'Impact of population growth', *Science* vol. 171, issue 3977, 1971, pp. 1212–7. This was the original publication of the formula.

3. An annual celebration of the state of being single, held on 11/11 each year. Singles Day, 2017, was the biggest one-day shopping event ever recorded and was widely reported in the financial and popular press.

4. Deloitte, *Voice of Asia: A rich tapestry of insights*, Deloitte University Press, Jan. 2017, accessed 2 May, 2017, https://dupress.deloitte.com/dup-us-en/economy/voice-of-asia/january-2017/insights.html. The extracts are from an upbeat assessment of Asia's economic prospects.

5. United Nations, *Report of the United Nations Conference on the Human Environment,* Stockholm, 5–16 June 1972, pp. 8, 52, accessed 2 May 2017, http://www.un-documents.net/ aconf48-14r1.pdf
6. *National Population Policy of Ethiopia,* Office of the Prime Minister, Addis Ababa, 1993, accessed 6 Dec. 2016, http://www.un.org/popin/regional/africa/ethiopia/policy/index.htm
7. The population policy document refers to an estimated annual loss of two billion cubic metres of soil from the highlands but other estimates are closer to one billion.
8. Shegena Zewdu, K.V. Suryabhagavan and M. Balakrishnan, 'Geo-spatial approach for soil salinity mapping in Sego Irrigation Farm, South Ethiopia', *Journal of the Saudi Society of Agricultural Sciences*, vol. 16, 2015, pp 17–24. Accessed 19 May 2017, https://doi.org/10.1016/j.jssas.2014.12.003
9. United Nations, Dept of Economic and Social Affairs, *World Population Prospects: The 2017 Revision, Key Findings and Advance Tables*, New York 2017, Table S3, p. 29, accessed 3 July 2017, https://esa.un.org/unpd/wpp/Publications/Files/WPP2017_KeyFindings.pdf
10. Sisay Nune Hailemariam, T. Soromessa and D. Teketay, 'Institutional arrangements and management of environmental resources in Ethiopia', *Environment and Natural Resources Research*; vol. 6, no. 1, 2016, pp. 67–87, http://dx.doi.org/10.5539/enrr.v6n1p67
11. International Development Partnerships (IDP) *Ethiopia - Environment Project.* Report, accessed 4 May 2017, http://www.idpuk.org/OurProjects/Environment/EnvironmentProject.htm
12. Tom Burgis, 'The great land rush' *Financial Times*, 2 March 2016, p. 9.
13. Norman Borlaug, on the occasion of the award of the Nobel Peace Prize in Oslo, Dec. 10 1970. Accessed 9 June 2017, http://www.nobelprize.org/nobel_prizes/peace/laureates/ 1970/borlaug-acceptance.html
14. Rockefeller Archive Center, 'Research institutes', accessed 8 June, 2017, http://rockefeller100.org/exhibits/show/agriculture/research-institutes
15. Prabhu Pingali and T. Raney, 'From the green revolution to the gene revolution: how will the poor fare?', ESA Working Paper No.

05-09, FAO, Rome, Nov. 2005. Accessed 9 June, 2017, http://www.fao.org/3/a-af276t.pdf

16. Prabhu Pingali, 'Green revolution: impacts, limits and the path ahead', *Proceedings of the National Academy of Sciences*, vol. 109, no. 31, 2012, pp. 12302–08, doi:10.1073/pnast0912953109
17. Jonathan A. Foley et al., 'Global Consequences of Land Use', *Science*, vol. 309, issue 5734, 2005, pp. 570–4, doi:10.1126/science.1111772. Note that land under trees grown for wood, and land abandoned in shifting cultivation, is excluded from the FAO definition of agricultural land.
18. FAO, *Aquastat – FAO's information system on water and agriculture*, accessed 5 July 2017, http://www.fao.org/nr/water/aquastat/water_use/index.stm
    The average of nearly 70% of water use being for agriculture is strongly biased by a few countries that have very high withdrawals.
19. FAO, 'Fertilizer use to surpass 200 million tonnes in 2018' media release, 16 Feb. 2015, accessed 10 June, 2017, http://www.fao.org/news/story/en/item/277488/icode
20. NASA's Gravity Recovery and Climate Experiment (GRACE) tracks groundwater changes around the world. Accessed 12 June 2017, https://grace.jpl.nasa.gov/applications/groundwater/
21. Manzoor Qadir et al., 'Economics of salt-induced land degradation and restoration', *Natural Resources Forum,* vol. 38, issue 4, 2014, pp. 235–307, doi:10.1111/1477-8947.12054
22. European Commission, 'Introduction to the new EU water framework directive', accessed 5 July 2017, http://ec.europa.eu/environment/water/water-framework/info/intro_en.htm
23. Brian Moss, 'Water pollution by agriculture', *Philosophical Transactions of the Royal Society B,* vol. 363, no. 1491, 2008, pp. 659–66, doi: 10.1098/rstb.2007.2176
24. Nikolaos Voulvoulis, K.D. Arpon and T. Giakoumis, 'The EU Water Framework Directive: from great expectations to problems with implementation' *Science of the Total Environment,* vol. 575, 2017, pp. 358–66, http://dx.doi.org/10.1016/j.scitotenv.2016.09.228
25. Robert Paarlberg, *Food Politics: What Everyone Needs to Know,* 2nd edition, Oxford University Press, New York, 2013, p. 73.
26. United Nations, *Report of the International Conference on Population and Development,* Cairo, 5-13 September 1994. Accessed 30 May 2017, https://www.ipci2014.org/sites/ipci2014.org/files/icpd_eng.pdf

27. United Nations, *Framework of Actions for the follow-up to the Programme of Action of the International Conference on Population and Development Beyond 2014.* Accessed 30 May 2017, https://www.unfpa.org/publications/framework-actions-follow-programme-action-international-conference-population-and
28. U.N., Millennium Development Goals, accessed 3 May 2018 https://unstats.un.org/unsd/mi/pdf/mdglist.pdf
29. United Nations, *Transforming our world: the 2030 Agenda for Sustainable Development.* Accessed 17 May 2017, https://sustainabledevelopment.un.org/post2015/transformingourworld

Chapter 19 Measuring up for sustainable development

1. Christopher J. Hoban and K. Tsunokawa, *Roads and the Environment: a Handbook,* World Bank, Washington, D.C., 1997, p. 63.
2. An overview of the MA, accessed 2 May 2018, can be found at: http://www.millenniumassessment.org/en/About.html
3. A guide to the MA reports, accessed 2 May 2018, can be found at: http://www.millenniumassessment.org/en/index.html
4. Millennium Ecosystem Assessment, *Ecosystems and Human Well-being: Synthesis,* Island Press, Washington, D.C., 2005, p.1.
5. Richard B. Norgaard, 'Finding Hope in the Millennium Ecosystem Assessment', *Conservation Biology,* vol. 22, issue 4, 2008, pp. 862–9.
6. WWF, *Living Planet Report 2018. Aiming Higher,* eds., M. Grooten and R.E.A Almond, WWF International, Gland, Switzerland. Accessed 31 Oct 2018, https://www.worldwildlife.org/pages/living-planet-report-2018
7. The major vegetation types (forest, grassland, tundra, desert) together with seas and the main freshwater bodies are called 'biomes' but the word is used loosely. For global conservation planning, the WWF has devised a classification system based on 26 'habitat types' subdivided into more than 800 'ecoregions' on land. In this scheme there are also more than 400 freshwater, and 200 marine ecoregions.
8. Erik Gómez-Baggethun et al., 'The history of ecosystem services in economic theory and practice: From early notions to markets and payment schemes', *Ecological Economics,* vol. 69, issue 6, 2010, pp. 1209–18, https://doi.org/10.1016/j.ecolecon.2009.11.007

9. Rudolf de Groot et al., 'Global estimates of the value of ecosystems and their services in monetary units', *Ecosystem Services*, vol. 1, 2012, pp. 50–61, https://doi.org/10.1016/j.ecoser.2012.07.005
10. Sharachchandra Lele et al., 'Ecosystem services: origins, contributions, pitfalls, and alternatives', *Conservation and Society,* vol 11, no. 4, 2013, pp. 343–58.
11. Millennium Ecosystem Assessment, *Ecosystems and Human Well-being: Biodiversity Synthesis,* Island Press, Washington, D.C., 2005, p.1. Accessed 14 June 2017, http://www.millenniumassessment.org/documents/document.354.aspx.pdf
12. Katie McShane, 'Is biodiversity intrinsically valuable (and what might that mean?)', in *The Routledge Handbook of the Philosophy of Biodiversity*, eds., J. Garson, A. Plutynski and S. Sarkar, pp. 155–67, Routledge, New York, 2017.
13. Walter E. Westman, 'How much are nature's services worth?', *Science*, vol. 197, issue 4307, 1977, pp. 960–64.
14. Robert Costanza et al., 'The value of the world's ecosystem services and natural capital', *Nature*, vol. 387, issue 6630, 1997, pp. 253–260.
15. Roy Haines-Young and M. Potschin , 'The links between biodiversity, ecosystem services and human well-being', in *Ecosystem Ecology: A New Synthesis (Ecological Reviews)*, eds., D. G. Raffaelli and C. L. J. Frid, pp. 110–39, Cambridge University Press, Cambridge, 2010.
16. Roy Haines-Young and M. Potschin, *Common International Classification of Ecosystem Services (CICES).* Consultation on Version 4, Report to the European Environment Agency, revised 2013. Accessed 9 July 2017, https://unstats.un.org/unsd/envaccounting/seearev/GCComments/ CICES_Report.pdf
17. In 2007, for example, following a meeting of environment ministers from the G8+5 countries in Potsdam, Germany, a global study was launched by the German Federal Ministry for the Environment and the European Commission. The study, now in its implementation phase, is called The Economics of Ecosystems and Biodiversity (TEEB). See: http://www.teebweb.org/about/
18. Roy Haines-Young and M. Potschin, Categorisation Systems: The Classification Challenge, Working Paper No. 15, Centre for Environmental Management, Nottingham University, 2016. Accessed 28 June, 2017, https://unstats.un.org/unsd/envaccounting/seearev/GCComments/CICES_Report.pdf

19. Alessandra La Notte et al., 'Ecosystem services classification: a systems ecology perspective of the cascade framework', *Ecological Indicators*, vol. 74, 2017, pp. 392–402, http://dx.doi.org/10.1016/j.ecolind.2016.11.030
20. US Department of Commerce, Bureau of Economic Analysis, 'How are GDP and related income measures of the national accounts affected by a disaster?' This faq is answered at: https://www.bea.gov/faq/index.cfm?faq_id=55
21. Richard T. Carson, 'Contingent valuation: a practical alternative when prices aren't available', *Journal of Economic Perspectives,* vol. 26, no. 4, 2012, pp. 27–42, http://dx.doi.org/10.1257/jep.26.4.27
22. The SNA, 2008, is produced and released under the auspices of the United Nations, the European Commission, the Organization for Economic Co-operation and Development, the International Monetary Fund and the World Bank Group. Accessed 2 May 2018, https://unstats.un.org/unsd/nationalaccount/sna2008.asp
23. Production of the SEEA Central Framework involved the UN Food and Agriculture Organization in addition to the 5 organizations responsible for the SNA. It is available from: https://unstats.un.org/unsd/envaccounting/seeaRev/SEEA_CF_Final_en.pdf
24. The SEEA-Experimental Ecosystem Accounting was the result of wide consultation with statisticians, economists, scientists and others from around the world. It is available from: https://unstats.un.org/unsd/envaccounting/eea_white_cover.pdf
25. From OECD Data: https:, the //data.oecd.org/gdp/gross-domestic-product-gdp.htm
26. Karl Falkenberg, 'Sustainability Now! A European Vision for Sustainability', European Political Strategy Centre, *Strategic Notes*, issue 18, 20 July 2016, p. 8. Accessed 11 July 2017, https://ec.europa.eu/psc/file/strategic-note-18-sustainability-now_en
27. Queensland Government, Department of State Development, *Carmichael Coal Mine and Rail Project: Project Overview*. Accessed 2 July 2017, https://www.statedevelopment.qld.gov.au/ assessments-and approvals/ carmichael-coal-mine-and-rail-project.html
28. Deloitte Access Economics, *At What Price? The Economic, Social and Icon Value of the Great Barrier Reef*, Brisbane, 2017. In this report the MA categories of ecosystem services are described, and the 'willingness to pay' method was used for contingent valuation. Accessed 9 July 2017, https://www2.deloitte.com/content/dam/

Deloitte/au/Documents/Economics/ deloitte-au-economics-great-barrier-reef-230617.pdf

29. UNESCO, *Great Barrier Reef – UNESCO World Heritage Centre*, accessed 10 July 2017, http://whc.unesco.org/en/list/154

Chapter 20 Growth

1. Leith van Onselen, 'Debunking Australia's fake growth record', *Macrobusiness*, accessed 11 Nov. 2017, https://www.macrobusiness.com.au/2017/06/debunking-australias-fake growth-record/
2. Mark O'Connor and W.J. Lines, *Overloading Australia: how governments and media dither and deny on population,* Envirobook, Canterbury, NSW, Australia, 2008.
3. Dietrich Vollrath, 'A brief history of economic growth', *Perspectives*, King's College London, 2015 Accessed 5 Aug. 2017, http://kcleconomics.com/a-brief-history-of-economic-growth/
4. M. Ayhan Kose and Ezgi O. Ozturk, 'A world of change', *Finance and Development*, (IMF) vol. 51, no. 3, 2014. Accessed 26 Aug. 2017, http://www.imf.org/external/pubs/ft/fandd/2014/09/kose.htm
5. IMF, *World Economic Outlook*, Apr. 2017, Chapter 2, http://www.imf.org/en/Publications/WEO/Issues/2017/04/04/world-economic-outlook-april-2017#Chapter2
6. IMF, *World Economic Outlook, Update*, July 2017, http://www.imf.org/en/Publications/WEO/Issues/2017/07/07/world-economic-outlook-update-july-2017
7. Dietrich Vollrath, 'A brief history of economic growth'.
8. Daniel Kahneman, *Thinking Fast and Slow,* Farrar, Straus and Giroux, New York, 2011. This book summarizes Kahneman's research into the distinction between quick, emotional thinking and the more considered type of thought. In 2002 Kahneman won the Nobel Prize in Economics for his work in this field.
9. Paul M. Romer, 'Mathiness in the theory of economic growth', *American Economic Review: Papers & Proceedings* vol. 105, no 5, 2015, pp. 89–93, , http://dx.doi.org/10.1257/aer.p20151066

10 UNCTAD, *The Least Developed Countries Report 2016,* UNCTAD, Geneva, 2016, http://unctad.org/en/PublicationsLibrary/ldc2016_en.pdf

11. United Nations, *World Economic Situation and Prospects 2017,* UN, New York, 2017, p.157, https://www.un.org/development/desa/dpad/wp-content/uploads/sites/45/publication/2017wesp_full_en.pdf

12. UNDP, *UNDP and the Concept and Measurement of Poverty*, UN, New York, 2016.

13 United Nations, *World Economic Situation and Prospects 2017*, p.5.

14. IMF, *World Economic Outlook, Update*, July 2017.

15. Christine Lagarde, 'The role of emerging markets in a new global partnership for growth', speech given at the University of Maryland, February 4, 2016, accessed 1 Nov. 2017, https://www.imf.org/en/News/Articles/2015/09/28/04/53/sp020416

16. M. Ayhan Kose and Ezgi O. Ozturk, 'A world of change'.

17. Esuna Dugarova and Nergis Gülasan, *Global trends: Challenges and opportunities in the implementation of the sustainable development goals*, UNDP and UNRISD, a report presented at a seminar in New York, 8 Feb. 2017.

18. Laurence Chandy, N. Ledlie and V. Penciakova, *The Final Countdown: Prospects for Ending Extreme Poverty by 2030*, Policy Paper 2013-04, The Brookings Institution, Washington, D.C., 2013, https://www.brookings.edu/wp-content/uploads/2016/06/The_Final_Countdown.pdf

19. ILO, *Decent work and the 2030 Agenda for sustainable development.* Accessed 2 Nov. 2017, http://www.ilo.org/global/topics/sdg-2030/lang--en/index.htm

20. ILO, *World Employment Social Outlook: Trends 2017*, Geneva 2017.

21. Robert H. Wade, 'Great escapes and great divergences: growth, poverty, and income inequality on a global scale' in *The Oxford Handbook of the Social Science of Poverty*, eds. D. Brady and L.M. Burton, Oxford Handbooks Online, Oxford University Press, 2015, doi:10.1093/oxfordhb/9780199914050.013.23

22. Laurence Chandy and B. Seidel, *How much do we really know about inequality within countries around the world? Adjusting Gini coefficients for missing top incomes*, Brookings Institution, Feb. 17, 2017. Accessed 2 Nov. 2017, https://www.brookings.edu/opinions/how-much-do-we-really-know-about-inequality-within-countries-around-the-world/

23. Robert H. Wade, 'Great escapes and great divergences.'

24. Branko Milanovic, *Global Income Inequality by the Numbers: in History and Now*, Policy Research Working Paper 6259, World Bank, Washington, D.C., 2012.
25. In principle, purchasing power parity (ppp) is a currency exchange rate based on the price of a comparable basket of goods in different countries compared to the price in USA. *The Economist* records a 'Big Mac' index where the McDonald's hamburger represents the basket of goods. It is not meant to be taken very seriously as the hamburgers are not exactly the same product around the world. Units of different currencies at ppp, at a given time, are sometimes referred to as international dollars.
26. Branko Milanovic, *Global Income Inequality by the Numbers*, p. 8.
27. World Bank, Gini Index, https://data.worldbank.org/indicator/SI.POV.GINI?
28. Credit Suisse Research Institute, *Global Wealth Report* 2016, Zürich, Switzerland, 2016.
29. Oxfam, *An Economy for the 99%*,, accessed 9 Oct. 2017, https://www.oxfam.org/en/research/economy-99
30. The Club of Rome website: https://www.clubofrome.org/
31. Donella H. Meadows et al., *The Limits to Growth*, Universe Books, New York, 1972. A digital scan version is available at: http://www.donellameadows.org/wp-content/userfiles/Limits-to-Growth-digital-scan-version.pdf
32. Paul R. Ehrlich, *The Population Bomb*, Ballantine Books, New York, 1968.
33. Julian L. Simon, *The Ultimate Resource 2*, Princeton University Press, Princeton NJ, 1996, Preface, p. xxxii.
34. Ibid., pp. 5–6.
35. Ibid., p.12.
36. David Coleman and R. Rowthorn, 'Who's afraid of population decline? A critical examination of its consequences', *Population and Development Review,* vol 37 (Suppl 1), 2011, pp. 217–48.
37. Herman E. Daly, 'Ultimate Confusion: the Economics of Julian Simon', *Futures*, vol. 17, issue 5 1985, pp. 446–50.
38. Paul and Anne Ehrlich, *Betrayal of Science and Reason*, Island Press, Washington, D.C., 1996, pp. 100–4.
39. Donella H. Meadows. J. Randers and D. Meadows, *Limits to Growth: the 30-Year Update*, Chelsea Green Publishing, White River Junction, VT, 2004.

40. Graham Turner, *Is Global Collapse Imminent?*, MSSI Research Paper No. 4, Melbourne Sustainable Society Institute, The University of Melbourne, 2014.

41. Commission on Growth and Development. *The Growth Report: Strategies for Sustained Growth and Inclusive Development*. World Bank, Washington, D.C., 2008, pp. 11–2.

42. OECD Environment Programme, *Indicators to Measure Decoupling of Environmental Pressure from Economic Growth*. Executive Summary accessed 24 Oct. 2017, https://www.oecd.org/env/indicators-modelling-outlooks/1933638.pdf

43. Herman E. Daly, *Steady State Economics,* 2nd edition, Island Press, Washington, D.C., 1991, p.119.

44. By OECD definitions, absolute decoupling occurs when the environmentally relevant variable is stable or decreasing while the economic driving force is growing. Relative decoupling occurs when the growth rate of the environmentally relevant variable is positive, but less than the growth rate of the economic variable.

45. European Commission, *EU Resource Efficiency Scoreboard.*

46. Ibid., p. 37.

47. Tim Jackson, *Prosperity Without Growth: Foundations for the Economy of Tomorrow*, Routledge, London and New York, 2017, p. 52.

48. John Stuart Mill, *Principles of Political Economy with some of their Applications to Social Philosophy*, Book IV, Chapter VI, 1848, C.C. Little & J. Brown, 1848. Mill's *Principles* can be downloaded as a free ebook from Google Books.

49. Herman E. Daly, Ed., *Towards a Steady-state Economy*, W.H. Freeman, San Francisco, 1973.

50. The mission statement for CASSE can be found at: http://www.steadystate.org/meet/mission/

51. Ed. Crooks, *Financial Times,* Nov. 23, 2009.

52. Tim Jackson, *Prosperity Without Growth: Foundations for the Economy of Tomorrow*, p. 174.

53. Giacomo D'Alisa, F. Demaria and G. Kallis, eds., *Degrowth: a Vocabulary for a New Era*, Routledge, New York and London, 2015.

Chapter 21 Food production

1. FAO, *The Future of Food and Agriculture – Trends and Challenges*, FAO, Rome, 2017, p. 6.
2. U N Convention to Combat Desertification, *The Global Land Outlook*, first edition, Bonn, Germany, 2017, pp. 52, 11, 169.
3. FAO, *Save and Grow Maize, Rice and Wheat: A Guide to Sustainable Cereal Production*, FAO, Rome, 2016, p. 9.
4. FAO, *The Future of Food and Agriculture,* pp. 32–4, 39.
5. Sarah K. Lowder, J. Skoet and T. Raney, 'The number, size, and distribution of farms, smallholder farms, and family farms worldwide', FAO. Published by Elsevier as open access document in *World Development,* vol. 87, 2016, pp. 16–29, , https://doi.org/10.1016/j.worlddev.2015.10.041
6. Benjamin E. Graueb et al., 'The state of family farms in the world', FAO. Published by Elsevier as open access document in *World Development* vol. 87, pp. 1–15, 2016, http://dx.doi.org/10.1016/j.worlddev.2015.05.012
7. Jacques Marzin et al., *Study on Small-scale Family Farming in the Near East and North Africa Region*, FAO International Centre for Advanced Mediterranean Agronomic Studies, Cairo, 2017. Accessed 28 Feb. 2018, http://www.fao.org/3/b-i6436e.pdf
8. James M. MacDonald and R. Hoppe, *Large Family Farms Continue To Dominate U.S. Agricultural Production*, U. S. Dept of Agriculture, 2017. Accessed 27 Feb. 2018, https://www.ers.usda.gov/amber-waves/2017/march/large-family-farms-continue-to-dominate-us-agricultural-production/
9. European Commission, 'Farm structures', 2017, in *Facts and figures on EU agriculture and the CAP*, from website accessed 9 Feb. 2018, https://ec.europa.eu/agriculture/sites/agriculture/files/statistics/facts-figures/farm-structures.pdf
10. Income is defined as entrepreneurial income, the EU definition of which can be found at: http://ec.europa.eu/eurostat/statistics-explained/index.php/Glossary:Entrepreneurial_income_account
11. European Commission, 'Agricultural and farm income', *in Facts and figures on EU agriculture and the CAP*, p. 18. Accessed 9 Feb. 2018, https://ec.europa.eu/agriculture/sites/agriculture/files/statistics/facts-figures/agricultural-farm-income.pdf
12. Frank Viviano, 'How the Netherlands Feeds the World' *National Geographic* magazine, Sept. 2017. Accessed 28 Feb. 2018,

https://www.nationalgeographic.com/magazine/2017/09/holland-agriculture-sustainable-farming/

13. Pew Commission on Industrial Farm Animal Production, *Putting Meat on the Table: Industrial Farm Animal Production in America,* 2008. Accessed 1 March 2018, http://www.pewtrusts.org/en/research-and-analysis/reports/0001/01/01/putting-meat-on-the-table
14. Lidwien A. M. Smit and D. Heederik, 'Impacts of intensive livestock production on human health in densely populated regions', *Geohealth,* Wiley Online Library, 2017, doi:10.1002/2017GH000103
15. Thomas P. Van Boeckel et al., 'Reducing antimicrobial use in food animals', *Science,* vol. 357, issue 6358, 2017, pp. 1350–2, doi:10.1126/science.aao1495
16. U.S. Environmental Protection Agency, *Nutrient Pollution*, accessed 1 March 2018, https://www.epa.gov/nutrientpollution
17. Adam Skolnik, 'The CAFO industry's impact on the environment and public health', *Sierra Magazine*, March 2017. This is not an impartial account. It contains personal experiences and an impressive aerial view of pig slurry lagoons.
18. Netherlands Environmental Assessment Agency, *Evaluation: Manure and Fertilizers Act, 2016,* synthesis of a report to the Dutch Parliament, accessed 3 Feb. 2018, http://www.pbl.nl/en/publications/valuation-manure-and-fertilisers-act-2016-synthesis-report
19. Reg A. Watson et al., 'Global marine yield halved as fishing intensity redoubles' *Fish and Fisheries*, vol. 14, 2013, pp. 493–503, doi:10.1111/j.1467-2979.2012.00483.x
20. FAO, *The State of World Fisheries and Aquaculture*, Rome, 2016, p.172, accessed Feb. 15, 2018, http://www.fao.org/3/a-i5555e.pdf
21. FAO, *Small Ponds Make a Big Difference: Integrating Fish with Crop and Livestock Farming*, FAO, Rome, 2000. Accessed 17 Feb. 2008, http://www.fao.org/docrep/003/x7156e/x7156e00.htm
22. Konrad Turkowsk and A. Lirski, 'The economics of carp farms in Poland', *Acta Ichthyologica et Piscatoria,* vol. 40, no, 2, 2010, pp. 137–144, doi:10.3750/AIP2010.40.2.06
23. FAO, *The State of World Fisheries and Aquaculture*, Rome, 2016, pp. 22, 71.
24. FAO, *Aquaculture Newsletter*, no 57, Sept. 2017, http://www.fao.org/3/a-i7851e.pdf.

25. Roland Blomeyer, et al., *The Role of China in World Fisheries,* Report to the European Parliament, Brussels, 2012. Accessed 22 Feb. 2018, http.//www.europarl.europa.eu/studies
26. Andrew Jacobs, 'China's appetite pushes fisheries to the brink' *New York Times*, 30 Apr. 2017. Accessed 23 Feb. 2018, https://www.nytimes.com/2017/04/30/ world/asia/chinas-appetite-pushes-fisheries-to-the-brink.html
27. Greenpeace, *Research Report on China's Trash Fish Fisheries,* Greenpeace East-Asia, 2017. Accessed 20 Feb. 2018, 2017_greenpeace_studie_china_jungfische_aus_beifang_als_fischfutter_in_aquakulturen_engl.pdf
28. Ibid., p. 21.
29. Jillian P. Fry et al., 'Feed conversion efficiency in aquaculture: do we measure it correctly?' Environmental Research Letters, vol.13, 024017, 2018. Accessed 1 March 2018, http://iopscience.iop.org/article/10.1088/1748-9326/aaa273/meta
30. Marine Harvest, *Salmon Farming Industry Handbook*, 2017, accessed 21 Feb. 2018, http://marineharvest.com/globalassets/investors/handbook/salmon-industry-handbook-2017.pdf
31. Gau Fu Mau, 'As demand grows, salmon farms spring up in China and Korea', *Seafood Source*, 2 Dec. 2016. https://www.seafoodsource.com/news/aquaculture/as-demand-grows-salmon-farms-spring-up-in-china-and-korea
32. Wild salmon fillets are pink because they contain a carotenoid pigment called astaxanthin. The pigment is produced by algae and it passes into the salmon's food chain. The flesh of farmed salmon without this pigment would be whitish so the pigment is added to pelleted food. The effect can be seen in flamingos, which for the same reason are pink or red in the wild but may be white in zoos. Astaxanthin is expensive but the synthetic form is less expensive than that obtained from cultured algae.
33. Donna Page, 'Calls for independent review of Port Stephen's fish farm after 20,000 kingfish hit marine park' *Sydney Morning Herald*, 2 Feb. 2018.
34. Hjeltnes B, Bornø, et al., eds., *The Health Situation in Norwegian Aquaculture 2016,* Norwegian Veterinary Institute, 2017, p.16. English version accessed 2 March 2018, www.vetinst.no:fiskehelserapporten/Fiskehelser apporten 2016

35. Hexing Wang et al., 'Antibiotic residues in meat, milk and aquatic products in Shanghai and human exposure assessment', *Food Control,* no. 80, pp. 217–225, doi:10.1016/j.foodcont.2017.04.034
36. U.S. Food and Drug Administration, *Aquacultured Seafood,* accessed 27 Feb. 2019, https://www.fda.gov/food/populartopics/ucm518782.htm
37. William H. Gaze et al., 'Influence of humans on evolution and mobilization of environmental antibiotic resistome', *Emerging Infectious Diseases*, vol. 19, no. 7, 2013, doi: 10.3201/eid1907.120871
38. Three mechanisms of horizontal gene transfer are well known. Conjugation involves contact between two bacterial cells but some bacteria can take in fragments of DNA from their environment by a process called transformation. DNA can also be passed from one cell to another by 'transduction' in which the DNA is conveyed by viruses known as bacteriophages. A fourth possibility, less well understood, is based on the transfer of DNA in vesicles released by bacteria.
39. William H. Gaze et al., 'Influence of humans on evolution and mobilization of environmental antibiotic resistome'.

Chapter 22 A synthesis

1. John Kay, *Other People's Money: Masters of the Universe or Servants of the People,* Profile Books, London, 2015, p.2.
2. Ibid., p. 177.
3. Andrei Kirilenko et al., *The Flash Crash: The Impact of High Frequency Trading on an Electronic Market*. This is a report to the U. S. Trading and Securities Commission, produced in 2010 and released to the public in 2014. Accessed 21 Mch. 2018, http://ssrn.com/abstract=1686004
4. John Kay, *Other People's Money*, p. 2.
5. YouGov, Optimism survey results, 2015, accessed 21 Mch. 2018, https://d25d2506sfb94s.cloudfront.net/cumulus_uploads/document/z2knhgzguv/GB_Website.pdf
6. J. van Zanden et al., eds., *How Was Life?: Global Well-being since 1820*, OECD Publishing, Paris, 2014, http://dx.doi.org/10.1787/9789264214262-en
7. J. Helliwell, R. Layard and J. Sachs, eds., *World Happiness Report 2017*, Sustainable Development Solutions Network, New York, 2017. Accessed 9 Dec. 2017, http://worldhappiness.report/ed/2017

8. John F. Helliwell, H. Huang and S. Wang, 'Social foundations of world happiness' in *World Happiness Report 2017*, pp. 8–47.
9. The top 10 were: Norway, Denmark, Iceland, Switzerland, Finland, Netherlands, Canada, New Zealand, Australia and Sweden.
10. Andrew E. Clarke et al., 'The key determinants of happiness and misery' in *World Happiness Report 2017*, pp. 122–43.
11. Richard A. Easterlin, 'Happiness, growth and public policy' *Economic Enquiry*, vol 51, issue 1, 2013, pp. 1–15, doi:10.1111/j.1465-7295.2012.00505.x
12. Andrew E. Clarke et al., 'The key determinants of happiness and misery'.
13. WHO, *Depression and Other Common Mental Disorders: Global Health Estimates*, Geneva, 2017, accessed 20 Mch. 2018, http://apps.who.int/iris/bitstream/10665/254610/1/WHO-MSD-MER-2017.2-eng.pdf
14. WHO, *Obesity and Overweight*, Factsheet, WHO Media Centre, Oct. 2017. Accessed 21 Mch. 2018, www.who.int/mediacentre/factsheets/fs311/en/
15. S. Randal Voss, M. R. Woodcock, and L. Zambrano, 'A tale of two axolotls', *BioScience*, vol. 65, no. 12, 2015, pp. 1134–40.
16. Luis Zambrano et al., *Ambystoma mexicanum, IUCN Red List of threatened species*, 2010, http://dx.doi.org/10.2305/IUCN.UK.2010-2.RLTS.T1095A3229615.en
17. Secretariat of the Convention on Biological Diversity, *Strategic Plan for Biodiversity 2011–2020 and the Aichi Targets, https://www.cbd.int/doc/strategic-plan/2011-2020/ Aichi-Targets-EN.pdf*
18. *UNEP-WCMC and IUCN, Protected Planet Report 2016,* UNEP-WCMC and IUCN: Cambridge UK and Gland, Switzerland, p. 35. Accessed 19 Mch. 2018, https://www.protectedplanet.net/c/protected-planet-report-2016
19. Nigel Dudley et al., 'National parks with benefits: how protecting the planet's biodiversity also provides ecosystem services' *Solutions Journal*, vol. 2, issue 6, 2011, pp. 87–95.
20. Marcel Alers et al., *Reducing Threats to Protected Areas: Lessons from the Field,* UNDP & World Bank, New York and Washington D.C., 2007.
21. Peter A. Lindsey et al., 'Underperformance of African protected area networks and the case for new conservation models: insights

from Zambia', *PloS One*, 2014, https://doi.org/10.1371/journal.pone.0094109

22. Ian Craigie et al., 'Large mammal population declines in Africa's protected areas', *Biological Conservation*, vol. 143, issue 9, 2010, pp. 2221–8.
23. Samudra Gupta Kashyap, 'The politics of cleaning up Kaziranga', *Indian Express*, 27 Sept. 2016. Accessed 22 Mch. 2018, http://indianexpress.com/article/india/india-news-india/kaziranga-eviction-drive-assam-elections-bjp-sonowal-government-politics-dead-3051774/
24. Lauren Coad et al., 'Measuring impact of protected area management interventions: current and future use of the Global Database of Protected Area Management Effectiveness', *Philosophical Transactions of the Royal Society B*, 2015, doi:10.1098/rstb.2014.0281
25. Marina Xavier da Silva et al., 'Effectiveness of protected areas for biodiversity conservation: mammal occupancy patterns in the Iguaçu National Park, Brazil', *Journal for Nature Conservation,* vol 41, Feb. 2018, pp. 51–62.
26. Peter Alexander et al., 'Losses, inefficiencies and waste in the global food system', *Agricultural Systems*, vol 153, 2017, pp 190–200, https://doi.org/10.1016/j.agsy.2017.01.014
27. Anne Mottet et al., 'Livestock: On our plates or eating at our table? A new analysis of the feed/food debate', *Global Food Security*, vol 14, Sept. 2017, pp. 1–8, https://doi.org/10.1016/j.gfs.2017.01.001
28. FAO, *The Future of Food and Agriculture – Trends and Challenges*, FAO, Rome, 2017.
29. Ian Heap and S.O. Duke, 'Overview of glyphosate-resistant weeds worldwide', *Pest Management Science*, 29 Nov. 2017, https://doi.org/10.1002/ps.4760
30. Natasha Gilbert, 'African agriculture: dirt poor' *Nature*, vol. 483, issue 7391, 2012, pp. 525–7. See also the Editorial of this issue, p. 510, doi:10.1038/483525a
31. Cameron M. Pittelkow et al., 'Productivity limits and the principles of conservation agriculture' *Nature,* vol. 517, issue 7534, 2015, pp. 365–8, doi:10.1038/nature13809
32. Ken E. Giller et al., 'Beyond conservation agriculture', *Frontiers in Plant Science*, vol. 6, no. 870, 28 Oct. 2015, doi:10.3389/fpls.2015.00870

33. David Cyranosky, 'Millions of Chinese farmers reap benefits of huge crop experiment', *Nature* (online), 7 Mch. 2018, https://www.nature.com/articles/d41586-018-02792-7
34. FAO, *The Future of Food and Agriculture – Trends and Challenges*, p. 52.
35. Ibid., p. 53.
36. Shell, *World Energy Model: a View to 2100*, 2017. Accessed 22 Mch. 2018, https://www.shell.com/energy-and-innovation/the-energy-future/scenarios/shell-scenarios-energy-models/world-energy-model.html
37. Jamie Condliffe, 'Clean coal's flagship project has failed' MIT *Energy Review*, 29 June 2017, accessed 5 Apr. 2018, https://www.technologyreview.com/s/608191/clean-coals-flagship-project-has-failed/
38. U.S Energy Information Administration, *International Energy Outlook, 2017*, accessed 20 Mch. 2018, https://www.eia.gov/outlooks/ieo/exec_summ.php
39. The sun's energy can be converted to electricity by two methods. With concentrated solar power (CSP), reflectors are used to heat up solutions from which steam is used to drive turbines as with other fuels. The more familiar method uses photovoltaic (PV) panels in which photons of sunlight displace electrons from a semiconductor material to create an electric current.
40. James Conca, 'EROI—a tool to predict the best energy mix', *Forbes* online posting, accessed 23 Mch. 2018, https://www.forbes.com/sites/jamesconca/2015/02/11/eroi-a-tool-to-predict-the-best-energy-mix/#39422dfea027
41. Jessica G. Lambert, C.A.S. Hall and S. Balogh, *EROI of Global Energy Resources: Status, Trends and Social Implications*, a report to the UK Dept. of International Devt. by the State University of New York, Oct. 2013.
42. Union of Concerned Scientists, *How Hydroelectric Energy Works*. Accessed 23 Mch. 2018, https://www.ucsusa.org/clean_energy/our-energy-choices/renewable-energy/how-hydroelectric-energy.html#.WrQqxC5uaiQ
43. Jessica G. Lambert, C.A.S. Hall and S. Balogh, *EROI of Global Energy Resources*, p. 65.
44. World Nuclear Association, *Energy Return on Investment*, 2017, accessed 3 Mch. 2018, http://www.world-nuclear.org/

information-library/energy-and-the-environment/energy-return-on-investment.aspx

45. S. Teske et al., *100% Renewable Energy for Australia – Decarbonising Australia's Energy Sector within one Generation.* Report prepared by the Institute for Sustainable Futures, Mch. 2016.
46. Michael Le Page, 'Europe's green energy policy is a disaster for the environment', *New Scientist*, 2 Dec. 2016. Accessed 22 Mch. 2018, https://www.newscientist.com/article/ 2114993-europes-green-energy-policy-is-a-disaster-for-the-environment/
47. Philip Fearnside, 'How a dam building boom is transforming the Brazilian Amazon', *Yale Environment 360*, Sept. 2017. Accessed 23 Mch. 2018, https://e360.yale.edu/features/ how-a-dam-building-boom-is-transforming-the-brazilian-amazon
48. Maslen Environmental, *Investigating the Impacts of Windfarm Development on Peatlands in England.* Report commissioned by Natural England, Apr. 2009. Obtainable from Natural England website: www.naturalengland.org.uk.
49. Daniel Nugent and B.K. Sovacool, 'Assessing the lifecycle greenhouse gas emissions from solar PV and wind energy: a critical meta-survey' *Energy Policy*, vol. 65, 2014, pp. 229–44, doi:10.1016/j.enpol.2013.10.048
50. National Academy of Sciences, *Renewable Fuel Standard: Potential Economic and Environmental Effects of U.S. Biofuel Policy,* National Academies Press, Washington D.C., 2011. Available from: http://nap.edu/13105
51. John DeCicco, 'Biofuels turn out to be a climate mistake – here's why', *The Conversation*, 5 Oct. 2016. Accessed 23 Mch. 2018, https://theconversation.com/biofuels-turn-out-to-be-a-climate-mistake-here's-why-64463
52. National Academy of Sciences, *Renewable Fuel Standard: Potential*, p.191.
53. Sarantis Michalopoulos, 'EU Parliament ends palm oil and caps crop-based biofuels at 2017 levels', *EURACTIV*, Jan. 19, 2018. Accessed 18 Mch. 2018, https://www.euractiv.com/ section/agriculture-food/news/eu-parliament-ends-palm-oil-and-caps-crop-based-biofuels-at-2017-levels/
54. Robert Engelman, 'Population and sustainability: can we avoid limiting the number of people?', *Scientific American*, June 2009.

Accessed 23 Mch. 2018, https://www.scientificamerican.com/article/population-and-sustainability/

55. United Nations, Dept of Economic and Social Affairs, *World Population Prospects: The 2017 Revision*, p.12.
56. Guttmacher Institute, *Adding It Up: Investing in Contraception and Maternal and Newborn Health, 2017*, Fact sheet. Accessed 23 Mch. 2018, https://www.guttmacher.org/ fact-sheet/adding-it-up-contraception-mnh-2017
57. David Coleman and R. Rowthorn, 'Who's afraid of population decline? A critical examination of its consequences', *Population and Development Review,* Vol. 37, supplement 1, 2011, pp. 217–48.

Index

Made in the USA
Monee, IL
04 November 2026

67381292R00203